Proceedings of the International Conference on

ADVANCES IN MANUFACTURING

9~11 OCTOBER 1984, SINGAPORE

An international event organised and sponsored jointly by IFS (Conferences) Ltd, Kempston, Bedford, UK and Singapore Exhibition Service Pte Ltd, Singapore

Co-sponsored by: The Singapore Robotic Association

Co-published by: IFS (Publications) Ltd and North-Holland (a division of Elsevier Science Publishers BV).

Proceedings of the
International Conference on
ADVANCES IN MANUFACTURING (AIM)

An international event organised and sponsored jointly by:

IFS (Conferences) Ltd, Kempston, Bedford, UK
and Singapore Exhibition Services Pte Ltd, Singapore

Co-sponsor: The Singapore Robotic Association

©October 1984 IFS (Conferences) and authors

Jointly published by
IFS (Publications) Ltd, UK
35-39 High Street, Kempston, Bedford MK42 7BT, England
ISBN 0-903608-58-8

North-Holland (a division of Elsevier Science Publishers BV)
PO Box 1991, 1000 BZ Amsterdam, The Netherlands
ISBN 0-444-87625-1

In the USA and Canada:
Elsevier Science Publishing Company, Inc.,
52 Vanderbilt Avenue, New York, NY 10017

Printed by: Cotswold Press Ltd, Oxford, UK

CONTENTS

Keynote

Industrial robot technology

Computer-aided design and planning

Computer-aided machining

Tooling and other aspects of advanced manufacturing technology

Advanced welding techniques

Computer-aided inspection and assembly

Late Papers

KEYNOTE

TRANSFER OF ADVANCED MANUFACTURING TECHNOLOGY INTO DEVELOPING COUNTRIES: PROBLEMS AND PROSPECTS FOR ACCELERATED DEVELOPMENT

I.S. Jawahir and W.C.K. Wong
Department of Mechanical Engineering
P.N.G. University of Technology
Lae
Papua New Guinea

ABSTRACT

This paper presents an overview of the manufacturing scene in developing countries. The focal point of the paper would be the transfer of advanced manufacturing technology, its adaptation and utilization for the establishment/expansion of export-oriented resource and/or demand based manufacturing industries in the developing world. A qualitative analysis covering aspects of compatibility in monetary potentials, material resources, network of local infrastructure and social and political factors is presented with emphasis on the need for an accelerated development strategy, directed towards higher productivity and better quality of manufactured products.

1. INTRODUCTION

It is difficult to explain the actual understanding of the word 'development', when applying this word to specify the levels of achievements of a country in a broad sense. However, it is understood that the words 'developing countries' or 'developing nations' and 'less developed countries' are very often referred to mean those countries other than developed countries and therefore this paper defines the words "developing countries" as the countries of the world excluding developed countries. The developing world of this category is understood to be consisting of about 75 percent of the total world population and is made up of a wide range of countries totalling almost 130 in number.

The most striking common features of the developing countries are such aspects as a low per capita income, a large rate of population growth and very often, a significant percentage of illiteracy. Almost all developing countries do have a need for self sufficiency to cope-up with the increasing rate of population growth. From the economic point of view, the most effective way of achieving this goal continues to be industrialisation. This would mean, improving the national economy through technological development and advancement. Each developing country does not follow an identical path to achieve this technology advancement mainly due to economic constraints and based on this, different priorities are set by each developing country. However, for most developing countries, the development of the manufacturing sector undoubtedly continues to be playing a vital role in building up the national economy.

In the aforesaid context, it is evident that for most developing countries, improving the existing manufacturing industry and/or establishing new manufacturing facilities, in general require the transfer of manufacturing technology from other countries. Considering the different levels of technology, currently existing in the developing countries, it becomes obvious that the rate and level of transfer of technology required varies from country to country. The manufacturing technology needed for a developing country is very often estimated through demand or import-substitution based manufacturing requirements [1]. Most developing countries, as yet, can not afford or do not anticipate the establishment of export-oriented or resource-based manufacturing industries on their own due to tighter economic constraints. The International Symposium on Science, Technology and Global Problems organised by COSTED in collaboration with UNESCO [2] very clearly indentified that, for a majority of developing countries transfer of manufacturing technology means

(a) The manufacturing process or processes available elsewhere is transferred to another place either in a modified form or otherwise to meet the specific requirements for the improvement of national economoy, or

(b) Indentification, adaptation and promotion of information on manufacturing activities which help to generate appropriate technologies indigenously.

The first definition deals with the import of manufacturing technology for use in the form of

(i) A turnkey project, or

(ii) Manufacturing products locally under a licence, or

(iii) Borrowing just the technical know-how with limited import of machinery, consultants, materials etc. and adapting it to meet the local condition.

The second definition is generally understood to be much broader and specifies the development strategy towards becoming technologically autonomous.

The choice of technology for a developing country may be based on a combination of the above mentioned categories both together. However, the proportion of high technology to indigenous technology will have to be appropriately worked out to suit each country individually.

The development trends in manufacturing technology obviously indicate the urgency and the need for an accelerated development strategy for developing countries to keep abreast with the continuously changing world of manufacturing needs. This situation inevitably requires a well balanced and compatible network of manfacturing activities in the developing world.

2. THE MANUFACTURING INDUSTRY IN DEVELOPING COUNTRIES: A GENERAL OUTLOOK

2.1 Process Technology and Productivity

The contribution of the manufacturing sector to GDP, in most developing countries has been increasing continuously with the increased rate of productivity growth. The local demand for manufactured products in a developing country is normally considered to be the prime concern in establishing manufacturing enterprises. The process technology for manufacture available in a developing country is very often a 'standard' bought-out package, which may not well suit the required size of the local market. Increased maintenance cost due to lack of spares and local skills for regular servicing, and excessive machine idleness due to unbalanced man to machine time scheduling are some of the disadvantages of such a standard and conventional technology, which caters for a small local market only.

It could be well argued that the excessive labour power, which is very characteristic for many developing countries, could eventually reduce the cost of production. However, a number of developing countries are confronted with increasing higher labour costs, which

2

are generally caused by various specific factors such as the concept of economic development of a given country and the socio-political factors behind the aforesaid system of economy. This situation tends to increase the cost of production, if the volume of production was to be restricted to a small-local market only. Improving the process technology available within the country could reduce the manufacturing costs to a great extent. This would mean that the transfer of more productive advanced manufacturing technology could be the solution for countries of this group for achieving higher productivity. Any form of advanced manufacturing technology would not merely be regarded as a means of achieving higher productivity. The type of advanced manufacturing technology needed is chosen according to the projected needs of a country. The selectivity of such appropriate advanced technology is a characteristic feature of almost every developing country [3].

2.2   Import Substitution Concept

Most developing countries have more or less accepted a development strategy to lead the economy through the concept of import substitution. This invariably requires boosting the local manufacture with a rapid programme purely on the basis of existing or projected demands. In achieving this target, many countries inevitably keep upgrading the level of locally available manufacturing technology and the skills. As the local demand grows with the increasing population, the manufacturing output must be raised. Increasing the manufacturing output would primarily require modernising the plant and equipment. This invariably means that the advancement of manufacturing technology would be the main strategic point to be considered for the concept of import substitution [4].

2.3   Resource-Based Major Industrial Expansion

Development of the manufacturing industry through resource-based establishment of major production facilities in developing countries has been a capital intensive exercise, which normally requires high technology and skills. Collaborative projects, forming a part of local semi-processing and manufacturing schemes are mostly funded either fully or partially by the collaborating country or an agency. This situation generally opens its doors to advanced manufacturing/processing technology.

2.4   Compatibility Requirements

Demand and resource-based manufacture in developing countries, while generally encouraging the introduction of advanced manufacturing technology, does require that the following compatibility requirements be satisfied before the process of transfer could be monitored [2].

(a)   Monetary Requirements

Perhaps this is the most important factor, restricting the transfer process in developing countries. In practice, many capital-intensive manufacturing technologies are transferred into developing countries either through foreign aid or as part of a contract with a multi-national company. Skills required are normally imported for the initial period of the project. Local demand alone would not be sufficient to justify the cost of such expensive ventures. Therefore, in importing such technology and skills, the export of manufactured products would have to be anticipated to make up the overall benefits of the venture.

(b)   Materials Requirements

Most developing countries have a common problem of being unable to supply suitably processed raw materials for the local manufacture of products. Imported technology will obviously be requiring the import of such materials from abroad. This is very often found to be not only time consuming but also expensive. Therefore the materials processing industrial development process will have to precede the transfer of technology process at some level so as to reduce the cost of manufacture.

(c)   Manufacturing Scale Requirements

Recent developments in the manufacture of advanced production machinery have repeatedly been proving the possibility of employing a range of machinery and equipment for a low or medium scale manufacture. The application of a computer aided work planning system developed to suit small and middle scale companies [5] is an added contribution towards improving productivity. The availability of a more productive technology for the specific manufacturing requirements of many developing countries will obviously be an encouraging factor to consider in planning for the immediate and projected needs of these countries.

(d)   Infrastructure Requirement

The developing countries, accepting the transfer of advanced manufacturing technology in the form of a turnkey project through an aid programme or otherwise will have a need for bearing the responsibility to maintain the manufacturing establishments in working order when the skilled expatriates leave these countries. Proper scientific and technological infrastructure would then have to be made available to provide services to the nation with its growing demand for manufactured products. It is therefore very essential for a developing country to train and educate the national workforce for facing the challenge of changing technology.

(e)   Social and Political Requirement

It is undoubtedly known that the industrial development policies in developing countries, being very specific to the particular country are formulated so as to accommodate the social factors such as the habits, customs and beliefs. It is also fairly obvious that other factors such as geographic location, marketability of manufactured products, transportation facilities etc. will additionally contribute to the setting up of industrial policies for any developing country. Developing countries, while looking forward to modernising the manufacturing facilities, have adopted various combinations of import substitution and export promotion policies. For many developing countries with a small domestic market, import substitution as a manufacturing advancement strategy has its limits; export-orientation, on the other hand, can provide a basis for sustained industrialisation. The governments of many developing countries welcome and encourage joint ventures in the manufacture of goods, while some restrictions are imposed on foreign exchange policies.

2.5   Production Planning Strategy

The strategy for planning production operations in developed countries is based on reducing the idleness of labour, which is normally very expensive. However, developing countries while not following an identical path with the development process of a developed country, would be required to plan the production activities through an appropriate analysis of the proportion between the machine cost and labour cost. Further, machine maintenance cost to machine cost ratio in most developed countries is relatively low mainly due to continuously changing innovative technlogies, which provide the total replacement of an existing machine after a certain period of operation, bringing the productivity rate higher. This not being the case in a developing country, the total maintenance cost spent on each machine time to time very often exceeds the cost of the machine, although this bitter fact is not fully understood clearly. Therefore, in planning production operations in developing countries the following factors must be considered.

(a)   Scales of production and economics.
(b)   Demand stimulations.
(c)   Maintenance requirements.
(d)   Machine idleness.
(e)   Labour costs.

An integrated approach to the specific situation would be required for formulating the manufacturing strategies for developing countries. The process of transfer of advanced manufacturing technology will have to be selected and adapted to suit the specific requirements for the projected future of a country. Planning the manufacturing facility for low volume markets has been considered to be encouragingly generating high profits in any given specialised area of manufacturing [6].

3.   MEANS OF TECHNOLOGY TRANSFER IN MANUFACTURING

The concept of employing high technology may have its limitations in some or in a group

4

of developing countries in the socio-economic context. It is however, well understood that the developing countries with relatively higher labour costs, requiring far more urgent needs of overcoming the constraints of development, would have to depend on some form of import of more productive manufacturing technology, which in turn could meet the challenge of the present and future requirements. The most common means of transferring advanced manufacturing technology for the developing world of this group would be as follows:

3.1   Direct Import

The technology developed, tested and successfully used in a developed country is directly transferred into a developing country normally in the form of a company to company transfer. The parent company or despatching company abroad will export this technology in the original form as it was introduced without any change or modification for the appropriate adaptation in the country of import. Atleast in a majority of developing countries this being the case, a number of inter-linked problems may be caused by this process of transfer. This includes:

(i)   Levels and size of manufacture for the importing developing country being very often small, the machine idleness becomes more [6].
(ii)   Maintenance cost and delay could become a major setback for continued use of the machinery and equipment imported.
(iii)   Lack of or inadequate local technical skills for continuous operation.

3.2   International Agency Assisted Projects

Most developing countries of relatively lower GNP are assisted by international bodies for development such as UNIDO and World Bank. The policy is "intensification and expansion of currently available technology" has been successfully functioning in several developing countries in Asia, Africa and Latin America. A case study on the establishment of an Engineering Design and Development Centre for improving the local manufacture with such an aid programme has been well explained in a paper written by Schmidth, Krainov and Ham [8].

In addition, the UNIDO has been actively participating in helping the developing nations in introducing new technologies to best serve the country atleast since the mid sixties. This is very often done at the requests of the respective countries. In 1975 the UNIDO having identified the need for developing the CAD/CAM area to meet rapidly growing needs of developing countries, has been involved with a number of developing countries in a wide range of CAD/CAM activities [9]. Some of the goals being anticipated by the UNIDO, in assisting the developing nations are:

(a)   Identifying those developing countries, which would benefit by the transfer of CAD/CAM Technology.
(b)   Setting up of a committee for the promotion of CAD/CAM in those countries.
(c)   Assisting those developing nations in the successful adaptation of CAD/CAM technology.
(d)   Promoting exchange of CAD/CAM information and experience among the countries of the developing world.

Several developing countries, which were planning to introduce CAD/CAM Systems have identified the following common problems:

- High cost of CAD/CAM Systems.
- Small local market.
- Lack of proper education and training locally.
- Lack of standardisation.
- Lack of confidence and inadequate readiness of manufacturing industries for implementation.
- Rapidly changing development trends of CAD/CAM technology.

A survey on "The introduction of CAD/CAM" conducted in Finland [10], having assessed the local situation, concludes that the following corrective measures could be used for reducing or eliminating the above problems.

- Improving education and training facilities.
- Exchange of skills with other countries.
- Co-operation in System development.
- Establishment of a public software library.
- Organising a common centre for process promotional activities.

A very recently published report on World Survey of CAM [11] identifies the following divisions in the taxonomy of CAM Systems.

- Flexible manufacturing systems.
- Computer aided process planning.
- Computer aided or scheduling
- Robotics.

The basic factor which should be borne in mind when speaking about the introduction of CAD/CAM in developing countries is that the transfer process of CAD/CAM techniques and the strategies for the development of these processes are very dependent on the circumstances: national politics, laws and legislation, the level of existing technology in the industry, size of companies, market demand, economy of transfer etc. Very often what is suitable for one country, need not be the best for another country [12].

3.3   Adaptation of Imported Technology

The policy on adaptation of an imported technology normally involves in some factors, which include:

- The use of local raw materials.
- The skills available in the country.
- The infrastructure of the country.
- The scale of manufacture.
- The socio-economic constraints.
- The cost to benefit ratio.
- The know-how for adaptation process.

The process of adaptation of an imported technology does not prevent the development of suitable technologies locally. In fact the adaptation exercise is more likely to generate some ideas for the local development processes. However, the prospects and incentives available for the development of an advanced manufacturing technology indigenously seems to be not very encouraging for many developing countries due to involvement of high capital and the expertise for the process.

3.4   The Role of Universities, Training and R&D Institutions

Academic Teaching and Training

It is very obvious that the role of a University in the technological development of a developing country is very vital. The level of technology required for the country primarily has to be identified by the man-power planning agencies and the Universities and Technical Training Institutions assist the country in meeting this requirement. Apart from teaching, these academic institutions are also engaged in conducting appropriate research programmes, running short courses and providing consultancy services to the industry. Most academic institutions in the developing world seem to be developing close liaison with the manufacturing industry, where young Engineers and Technologists are given practical training. Many final year student projects are chosen from a wide range of industrial problems-based topics. The manufacturing industry also encourages short-term academic research, which involves in feasibility studies on the advancement of existing manufacturing facilities. Typical examples of such projects include:

- Improvement of process technology.
- Man-machine studies.
- Introduction of work study techniques.
- Improvement of plant-layout.

6

(e)   Computer applications in manufacture.
(f)   Low-cost automation.
(g)   Improvements in Quality Control Techniques.
(h)   Part-classification Systems and Introduction of Group-technology.

Research and Development

It is also very common that the research, conducted by the academics and researchers in developing countries are mostly funded by government agencies.

Limited funding, lack of adequate R&D policies are some of the most important characteristic features of a developing country.   High risks involved in the expenditure on R&D in a developing country is perhaps the main reason for the under-rated R&D activities in the developing world.   Manufacturing industries sponsoring limited R&D activities are normally not much interested in conducting long-term R&D work.

Training for Trade Skills

Despite manufacturing-oriented academic teaching and training in the developing world, an acute problem of paying insufficient attention to the training of technicians and tradesmen still exists [13].   Especially the requirements of skilled tool makers, fitters, machinists, etc. for the manufacturing industry is very often found to be in short supply.   Conventional manufacturing technology available in most developing countries results in idleness of machine and/or long machine shut-down time mainly due to above reasons.   Appropriately advanced manufacturing technology will not require skilled operating personnel, and the lack of skilled workers, as it exists, would no longer be a problem for the developing world.

3.5   The Role of Professional Scientists and Technologists

It is generally expected that the scientists and technologists, practicing their professions in the developing countries have normally acquired the competence to evaluate, select, and regulate the production operations.   They are also expected to acquire the capacity to assimilate and gradually improve the adopted technologies to suit the local condition. The following requirements are to be met by the scientists and technologists in developing countries [14].

(a)   Professional competence.
(b)   Science and technology planning on the basis of priorities.
(c)   Improved communications between Scientists/Technologists and Administrators.
(d)   Work freedom for Scientists/Technologists.
(e)   Non-governmental organisational activities.
(f)   Exchange of scientific information with other Scientists/Technologists (internally and internationally).

3.6   The Role of Manufacturing Industry

Labour intensive manufacturing technology is the most common form of technology found in many developing countries, where the labour costs are cheap.   However, the size of the market, prospects for export of manufactured products, increasing labour costs and overhead charges demand the modernisation of production machinery.   This being the case for a number of rapidly developing countries, opens gateways to the introduction of advanced manufactured techniques.   The following forms of transfer are found to be in common for most growing industries of developing nations.

(1)   Direct transfer under the sponsorship of a parent company.
(2)   Introductory concessions offered by the manufacturers of high technology machinery, through expositions, exhibitions, seminars organised by professional bodies or government and private sector organisations.
(3)   Modification of existing machinery for specific requirements using imported know-how.
(4)   Development of relevant function-oriented additional accessories/attachments locally for incorporating into the basic machinery/equipment.
(5)   Complete design and development of the machinery for high production scales.

The long-awaited economic recovery is now in progress. This recovery is accompanied by new demands on manufacturing establishments to increase productivity and to re-orient the manufacturing sector of a developing country to face an intensive competitive pressure from both domestic and foreign competition. Manufacturing industries are ripe for major restructuring. The adaptation and incorporation of new product and process technologies into the industry should prove to be an attractive strategy for many single and multi-plant companies [15].

4. GENERAL REMARKS

Notwithstanding the above remarks on the many problems confronting implementation of advanced manufacturing technology into developing countries, there remains on a brighter side many scopes for advancement.

To begin with, it is of our opinion that too much emphasis had been placed on import substitution. It is equally true that many developing countries simply adopt technology from other developed countries. These undertakings may appear worthy of attempts but most unfortunately the success rates have been very low and are particularly unsuitable for long-term plans. Instead, developing countries should take a positive attitude and concentrate on adapting or better still, developing modern technology for its exports utilising its own resources (natural, manpower etc.) and capabilities.

Customers of today not only demand on a variety of products but also short delivery time and are more quality conscious. Recent experience has shown that product life cycles have become much shorter. In developing countries, high quality products and productivity are still the two main issues of concern. Among the reasons for this lagging quality and productivity performance are the outdated, outmoded production methods and management techniques being used in industrial plants where capital investment has been minimal. A crude estimate of the equipment used in production, in fact, is at least about two to three decades old.

With cost and competition up and capital investment stagnant, manufacturers no longer can afford to drag on with 'bandaid' solutions. For manufacturing concerns to progress from organised chaos to productive factories of the future, it is vital that increased productivity and consistent quality (at acceptable level) be emphasised in long-term strategic and modernisation plans. To this end, governmental support is deemed essential. In Japan for instance, the government is directly involved in enhancing the nation's industrial base. Through a co-operative, government-industry-academic partnership, Japan is pushing the

state-of-the-art in a number of manufacturing areas including computer technology. In some instances, the original concept of totally automated factories is no longer a fiction but has become fruition. Similar successful examples involving government and industry-academic liasion can be sighted in West Germany, Norway, Taïwan and Singapore.

The era ahead of developing countries in manufacturing is one full of challenge and excitement. It is vitally important that manufacturing engineers have positive attitudes for work and possess the quality of being innovative and entrepreneurship. Advanced manufacturing technologies (such as FMS) would hold the promise to restore efficiency, simplify operations, maintain quality and increase productivity.

The status of computer-aided advanced manufacturing technology in Singapore with its applications in the areas of NC/CNC Machine Tools, Robots and Manupulators, Resource Planning, Flexible Manufacturing Systems, Computer-Aided Inspection and Computer-Aided Design, has been very extensively studied and presented in a paper by Yang [16]. It is also expected that further significant contribution to this effect will be made by various internationally known scientists and academics in the plenary sessions of this conference.

5. CONCLUSIONS

5.1 The need for accelerated development in the manufacturing sector of developing countries has been identified.

5.2 The accelerated development process through improved productivity could be achieved for a number of developing countries by transferring advanced manufacturing technology at appropriate levels and scales.

5.3   The proportion of high technology to appropriate indigenous technology required by a developing country would vary from country to country and hence the development strategy for each country is unique.

5.4   Increasing the rate of productivity growth and producing high quality products could be the two main strategic targets for long-term planning of the developing world. This would inevitably require the use of advanced manufacturing technology.

5.5   Projected trends in manufacturing demands bridging the gap between the countries of different economy through export-oriented and improved manufacturing process technologies and management.

5.6   The Universities, Technical Training Institutions, R&D organisations would be expected to take a lead in transferring more productive manufacturing technologies to suit the appropriate needs of developing countries.

REFERENCES

1.   "Manufacturing in Australia", Vol.2, Working Party Papers, The Task Force on Manufacturing, The Institution of Engineers, Australia, March 1980.

2.   "Technology Transfer and Industrialisation", (views from the developing world), Proc. of Int. Symposium held in Kuala Lumpur, Malaysia, organised by COSTED, 27-30 April 1979, pp.158-188.

3.   Kazem Behbehani and M.S. Marzouk, "Transfer of Technology and Selectivity of Appropriate Science and Technology", Proc. of Int. Symposium on Science and Technology for Development, Singapore, January 1979, pp.119-122.

4.   Jawahir, I.S., and Wong, W.C.K., "The State of Manufacturing Industry in Papua New Guinea", Paper presented at the Int. Conf. on Manufacturing Engineering, (ICME-83), Singapore, June 1983.

5.   Tonshoff, H.K., Ehrlich, H., Meyer, K.D., and Prack, K.W., "Development of a Computer Aided Work Planning System for Small and Middle Scale Companies", Proc. of Int. Conf. on Manufacturing Engineering, (ICME-80), Melbourne, Australia, August 1980, pp.262-267.

6.   Stamm, W.J., "Planning the Manufacturing Facility for Low Volume Markets", Paper presented at the Int. Conf. on Manufacturing Engineering, (ICME-83), Singapore, 1983.

7.   Mazhar Ali Khan Malik, "Technology Transfer in Production Planning for Developing Countries", Int. Journ. for Production Research, Vol.17, No.3, 1979, pp.259-264.

8.   Schmidth, A.O., Krainov, N.H., and Ham, I., "Manufacturing Engineering for Developing Countries", Proc. of Int. Conf. on Production Technology, Melbourne, Australia, 1974, pp.95-101.

9.   George P. Putman, "A Strategy for CAD/CAM Technology Transfer to Developing Countries", Proc. of the IFIP, WG5.2, Working Conf. on CAD/CAM as a Basis for the Development of Technology in Developing Nations, Sao Paulo, Brazil, Oct.21-23, 1981, pp.235-240.

10.   Uusitalo, M, Nykänen, M., and Kautto-Koivula, K., "Survey of CAD/CAM Situations in Finland", Technical Research Centre of Finland, Dec. 1980.

11.   Hatvany, J. (Editor), Merchant, M.E., Rathmill, K., and Yoshikava, "World Survey of CAM", Butterworths Publication, 1983, p.19.

12.   Kaisa Kautto-Koivula, "Strategies for CAD/CAM Technology Transfer", Example from

Finland. Proc. of the IFIP WG5.2, Working Conf. on CAD/CAM as a Basis for the Development of Technology in Developing Nations, Sao Paulo, Brazil, Oct. 21-23, 1981, pp.271-279.

13. Irvine, D.H., "Science and Technology in Developing Countries", Proc. of Int. Symposium on Science and Technology for Development, Singapore, January 1979, pp.99-104.

14. Innas Ali, M., "The Role of Scientists and Technologists in the Development of Less Developed Countries", Proc. of Int. Symposium on Science and Technology for Development, Singapore, Jan. 1979, pp.221-223.

15. McIntyre, B., "NC-CNC, The Future from the Users View-point", Paper presented at the Int. Conf. on Manufacturing Engineering, (ICME-83), Singapore, June 1983.

16. Yang, L.J., "Training in Computer-Aided Manufacturing for Mechanical and Production Engineers in Singapore", Proc. of the Regional Conf. on Recent Developments in Engineering Educ., Bangkok, Thailand, Dec. 12-16, 1983, pp.29-48.

INDUSTRIAL ROBOT TECHNOLOGY

'INDUSTRIAL ROBOTS IN SINGAPORE'

L J YANG
Senior Lecturer
School of Mechanical & Production Engineering
Nanyang Technological Institute
Singapore

ABSTRACT

A robot is capable of performing many tasks better than a human operator. It is also a key element in the modern flexible manufacturing system.

Since Singapore decided to move into high value added industries through increased levels of mechanisation and automation in 1979, a number of robots have been installed in Singapore industries.

There are now a total of 86 industrial robots installed in 37 companies. About 75% of these robots are using electrical drives and about 18% are using hydraulic drives.

Regarding the application of these industrial robots, about 20% each are used for material handling and spray printing; about 19% for educational training and about 17% each for assembly and for arc welding.

A survey conducted by the Economic Development Board has indicated that more companies are expected to install robots in the near future. The projected demand would be in the region of between 1,000 to 2,000 units within the next five years.

So far there is only one manufacturing company involved in the production of industrial robots in Singapore. The robots are of the pick and place type. Since we have a good base of electronic and supporting industries and a good demand for industrial robots, more manufacturing companies should venture into the development and manufacturing of robots and their controls in Singapore.

1. INTRODUCTION

The Robot Institute of America defines the robot as "a reprogrammable, multifunctional manipulator designed to move material, parts, tools or specialised devices through variable programmed motions for the performance of a variety of tasks".

A robot therefore must have inbuilt intelligence and has the ability to operate automatically. A robot is not capable of performing all the tasks that its human counterpart can achieve. However robots are able to do some jobs better than human operators. Robots also have many outstanding features when compared with human operators. Some of the examples are:

o A human operator has to leave for home after a shift period of between eight and twelve hours. However a robot can be programmed to work round the clock everyday and seven days a week if so required.

o A human operator has to take time off for eating and for other physical needs. Robots do not need time off for such reasons except very rarely breakdowns and maintenance.

o Strikes, go-slows, overtime bans and other disruptive action from the labour force can hit any production plant. A recent example is the West Germany's metal workers who demanded a 35-hour week. However a robot will always obey your command.

o Human operators get tired and may not be able to work efficiently when they are sick or under unpleasant working conditions. Robots however can operate in extreme conditions such as noise, heat, vibration, toxic fumes, smells, cramped workspace etc.

It can therefore be seen that the prime advantage a robot has over human labour is its productivity. However the main advantage a robot has over special purpose automation is its flexibility and the role it plays in modern manufacturing systems.

Until recently, the international competition has been based on mass production which depends on exploitation of economies of scale by using larger production volumes and greater standardization in order to cut costs and thus be able to compete on the basis of price. However the growing sophistication and diversification of demand for greater variety of goods from the increasing affluent consumers have altered the situation.

The future trend in manufacturing will be the flexible manufacturing systems where industrial robots play an important role in either the loading and unloading of component parts, inspection and measurement or the actual production and finishing operations.

Singapore decided to restructure its economic policy in 1979. Its main objectives are:

o to move into high value added industries through increased levels of mechanisation, automation and computerisation, and enhanced productivity.

o to develop a wholly Singaporean workforce by 1990 through a corrective wage policy and the gradual phasing out of non-local labours.

Productivity will grow as management and workers adopt better work methods and become more skilful. However the greater part of productivity growth will only come from manufacturing more high value added products and from the introduction of modern manufacturing technology.

To create a wholly Singaporean workforce, mechanisation and automation is necessary for processes that are hot, dirty or unpleasant and already shunned by local workers.

The introduction of robots in Singapore industries will therefore help to achieve the above mentioned objectives.

2. STATUS OF INDUSTRIAL ROBOTS IN SINGAPORE

Hydraulically operated manupulators were introduced into a forging plant about twelve years ago for transferring heated steel billets from a furnance onto the die cavity of a press. To date, there are more than 400 manupulators installed in about seventy (70) companies in Singapore for various industrial applications.

Industrial robots were however introduced into the Singapore industries after the Government's decision to restructure its economic policy in 1979. To date, the total number of industrial robots in use is 86 units. There are altogether 37 companies using industrial robots.

Fig. 1 shows the growth of industrial robots in Singapore.

Regarding the drive mechanism used, about 75% of the robots installed in Singapore are using electrical drives and about 18% are using hydraulic drives.

Regarding the application of these robots, about 20% each are used for material handling and spray painting, about 19% for educational training, and about 17% each for assembly and for arc welding.

Fig. 2 shows the application of industrial robots by Singapore industries.

From a survey of 1300 manufacturing companies conducted by the Economic Development Board in May 1983, it was found that:

o Another 180 companies are expected to install robots and manupulators over the next five years.

o The projected demand for manupulators and robots over the next five years is expected to range from about 1,000 - 2,000 units, about half of which are expected to be reprogrammable robots.

o Robots are expected to be used mainly in

   . metal working (19%)
   . inspection/measurement (19%)
   . assembly (15%)
   . plastics (15%)
   . welding (10%)
   . spray painting (5%)
   . casting (5%)

3. ROBOTIC INDUSTRIES IN SINGAPORE

Although presently there are many companies engaged in machine tool, industrial machine and computer peripheral manufacturing, there is only one involved in the production of industrial robots.

The robots which are based on an American design from Machine Dynamics Incorporated are electro-pneumatic controlled, programmable and applicable for pick and place functions.

The company is currently producing three types of industrial robots.

o The Robota 10A1 which is capable of handling workpieces of about 1 kg weight, is suitable for presses, injection moulding machines, die casting machines and assembly lines.

o The Robota 10B5 which is capable of lifting workpieces of 5 kg load, is suitable for use in the plastic and metal working industries.

o The Robota 10C10 which can handle workpieces of up to 10 kg weight, is designed for material handling operations in foundry, plastics and metal working industries.

Price has been a major deterent factor in moving companies to incorporate robots in their workforce. Locally produced industrial robots would definitely have the advantage of price and the flexibility of integration into existing production systems. By reducing the price, more companies will benefit from the installation of robots.

Robotics has been identified by most of the industrialised countries as one of the key industries in the 1980s and beyond. It is a high growth industry with potentials for expansion as new applications and markets are opened up. Other factors that would favour the development of robotic industry in Singaproe are:

o A good base of electronic component companies and supporting industries which can supply parts at prices lower than their European and American counterparts

o A young and well educated workforce who are more adaptable and receptive to new technology.

o Manufacturers will have to resort more to automation and industrial robots because of labour shortage.

It is therefore suggested that more companies should venture into the manufacturing of robots and their control elements, preferably the more advanced types.

4. TRAINING OF ENGINEERS IN ROBOTICS

To meet the manpower requirement of the new technology era, both the Economic Development Board and all the institutions of higher learning in Singapore have organised various courses on robotics.

The School of Mechanical and Production Engineering of Nanyang Technological Institute used a number of teaching robots during the in-house practical training period in February 1983. Students were given hands on experience on the programming and assembly of these robots. The equipment included two Eltec RB-2 robots, three Rhino XR-2 robots and the Feedback HRA 933 hydraulic robot.

In their 3rd year, some of the students were given the opportunity to design a robot arm and to study the control and programming of robots using the Basic language on microcomputers. Students will be exposed more on robotics in their final year course through a number of lectures and projects.

Two industrial robots were installed at the end of 1983 and early 1984. These are model IRB6 from ASEA, Sweden. One will be used for training the students on robotics in general while the other which is complete with a laser sensor will be used for training students in robotic arc welding.

The ASEA-EDB Robotics Training Unit (AERTU) was opened on December 1983. It was set up jointly by ASEA AB of Sweden and the Economic Development Board. It was establised to facilitate industrial automation in Singapore by providing training in robotic technology. The training programme involves theory and extensive hands on practice which will enable participants to operate, program and maintain robots. The following courses are available:

o Programming and operation of industrial robots.
o Project engineering for robotisation.
o Mechanical maintenance.
o Electronic maintenance.

The high percentage of industrial robots used for educational training as shown in Fig. 2 indicates the emphasis by both the EDB and Institutions of higher learning in training manpower required by our industries.

5. <u>CONCLUDING REMARKS</u>

The comparatively large number of robots installed indicates that Singapore industries have recognised the need for robotisation in order to stay competitive in the world market. The decision to install a robot for any particular application must of course be made after studying all the economic factors (eg. costs and benefits) as the success of an industrial undertaking has to be measured in terms of financial performance. The manufacturing of more robots in Singapore will definitely have added advantage to our objectives of improving productivity through more mechanisation and automation.

It should also be noted that majority of the robotic systems installed so far are of the conventional types which are restricted to repetitive tasks such as spray painting and machine loading. To enable robots to be easily taught to react and adapt to changing circumstances, intelligent vision sensing and tactile sensing devices must be incorporated into the robotic system. The Singapore manufacturing industries will benefit more if more R & D could be directed into this area.

<u>ACKNOWLEDGEMENT</u>

The author would like to thank Miss Jesamine Lim for typing the manuscript.

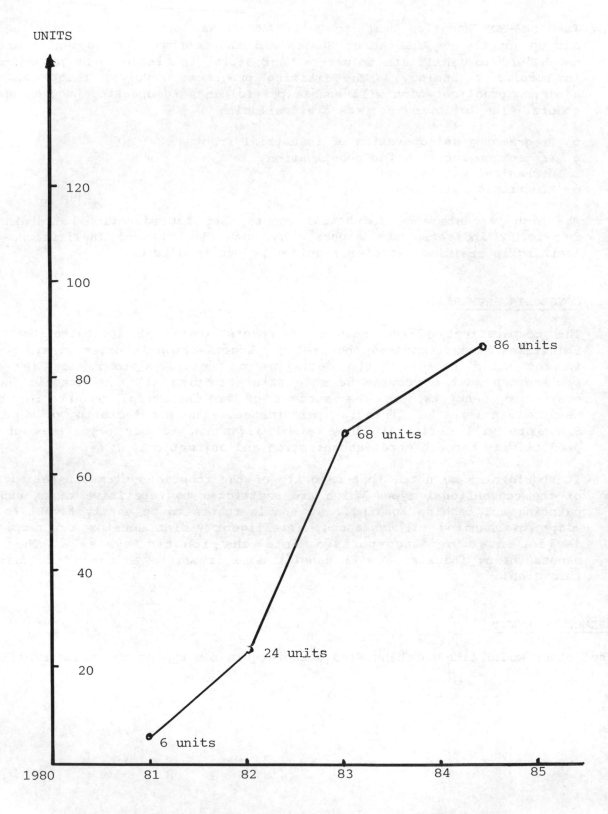

FIGURE 1   GROWTH OF INDUSTRIAL ROBOTS IN SINGAPORE

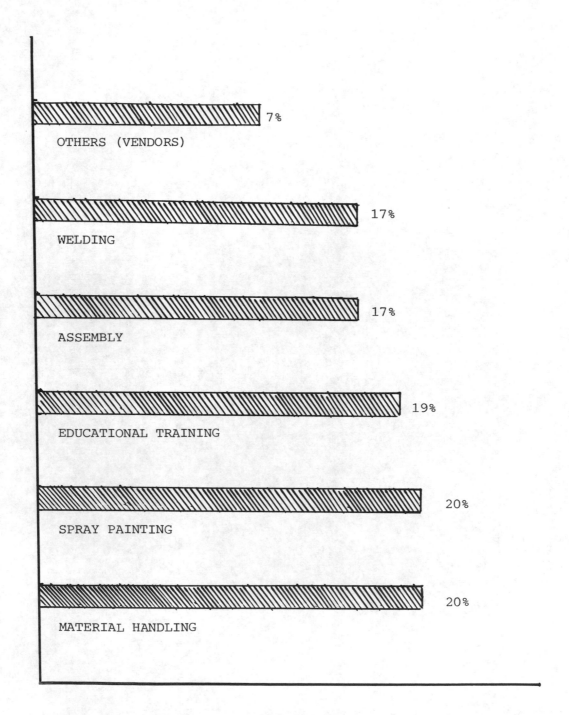

FIGURE 2   APPLICATION OF INDUSTRIAL ROBOTS IN SINGAPORE INDUSTRIES

APPLICATION OF INDUSTRIAL ROBOTS - POSSIBILITIES, CHANCES AND RESULTS

H.-J. Warnecke, R.D. Schraft, Fraunhofer-Institute for Manufacturing Engineering and Automation, (IPA), Stuttgart, Fed. Rep. of Germany

ABSTRACT

Industrial robots have been familiar in the industry since 1965. The first significant applications were in Europe in the early 70's and by the end of 1983 approx. 13,700 industrial robots were in use in Europe. The application areas can be divided into tool handling and workpiece handling. On examining the robot applications in the Federal Republic of Germany, one finds that approx. 50 % of the devices are applied in the automobile industry. Apart from the classic application areas such as spot welding, the loading of machines and arc welding, large numbers of new applications are expected in the future in the machining, loading and unloading of machines as well as the assembly fields.

The following report gives an overall view of the state-of-the-art, as well as of further potential application areas.

APPLICATION NUMBERS

Federal Republic of Germany

On examining the industrial robot market in the Federal Republic of Germany, one can see that by the end of 1983 approx. 200 different types were being offered by 80 firms. This figure is of importance when one considers the fact that the three largest manufacturers and distributors in the Federal Republic of Germany alone, cover 50 % of the market. The first dozen manufacturers and distributors in the Federal Republic of Germany already supply 80 % of the market. The industrial robot market is certainly a growing market and a chance for small and medium sized firms. In several industrial robot application fields there are already clear market leaders, whereas other areas, for example, the growing assembly field, still seem to have many possibilities open. The industrial robot however, can not and must not become the total cure for all automation problems of the future, but it has found a steadily increasing scope of applications in production.

International

The research and development of industrial robots has greatly influenced their utilization. This research and development work began sooner in the USA and Japan than in Europe. Not considering various definition questions, Japan is the leader of all countries using industrial robots (Figure 1). At the end of 1983 Japan had already installed 16,500 industrial robots, as opposed to 4,800 in the Federal Republic of Germany, which makes it clear that, in 1983, the japanese were still clearly in the lead against the americans with 8,000 and the europeans with 9,000 industrial robots. The latest trends in Japan have created high expectations of strong increases in the assembly field.

Absolute industrial robot application numbers have little meaning with regard to the readiness for innovation of the individual industrial nations. A better insight is obtained by considering the density (devices installed per 10,000 employed, Figure 2). Hereby it becomes clear that Sweden lies far ahead. The reasons for this are the strong social system, the high labour costs and the presence of potential users (automobile industry). In addition to this, efficient industrial robots were being offered at a very early stage on the scandinavian market.

APPLICATION AREAS

The following describes the most frequent industrial robot application areas. Figure 3 gives an overall view of the fields of application of industrial robots. The first differentiation to be made is between industrial robots for tool handling and for workpiece handling. If the industrial robots are equipped with a gripper, or respectively with a tool changing device, a combined application is possible.

The development of each application area shown in Figure 4 has been very different. The present state and the developments of the past years are presented in figure 4. The following comments may be made on the developments shown:

Coating

There are no essential, unsolved, technical problems. The economy alone decides on the application (Figure 5).

Spot Welding

The very strong increase of spot welding industrial robots is attributed to the suppressed demand of the automobile industry. This development is not expected to continue, since the satiation of the automobile industry can already be foreseen (Figure 6).

Arc Welding

Generally, very good prospects are seen for the future of arc welding, as soon as solutions are found for the tracking of welding seams, balancing of mounting tolerances and similar, as yet unsolved, problems (Figure 7).

Deburring

Deburring and machining with industrial robots is a future field of work which, due to many technical problems, is still in its early stages.

Assembly

A delphi-questionnaire (Figures 9 and 10) forecasts the overall highest growth rate for industrial robots in assembly. Not only the product structure, but also the technics of the device fulfil the prerequisites for a successful application in this field (Figure 11).

Workpiece Handling

The problem of workpiece handling has principally been solved. In particular the loading and unloading of machines will, in future, be of economic importance for the multiple shift operation of capital-intensive manufacturing devices (Figure 12).

Figure 13 shows an up-to-date forecast of the changes in industrial robot application fields between 1981 and 1991. The forecast is based on the USA.

Similar relative distributions may also be expected in Europe, however, as strong similarities were the result of a comparison of figures in 1981.

Experience to date has shown that the costs for engineering and peripherals and the costs for the handling device stand at a ratio of approx. 1:1. At present it is usually not in the interest of the user to buy an industrial robot in the form of automation components. The demand is for overall system solutions.

By the use of gripper changing devices, industrial robots are also able to take over complex tasks. Different tools as well as different workpieces may be handled at one industrial robot workplace (Figure 14).

DEVELOPMENT TRENDS

Looking further into the future, the following developments may be expected in robot technology:

1. Sensor technology will develop at a very high rate and will enable robots to adjust to the given task with more "feeling" and "intelligence".

2. Grippers will become more universal. Together with the sensor

technology, a hand-eye-coordination in industrial surrounding will also be possible.

3. The programming of industrial robots will be simplified by other programming methods (CAD/CAM, language input, higher programming languages, half-automatic programming and others). The programme development costs and time will therefore decrease.

4. The robots will be able to work in a larger area, i.e. they will become more mobile (<u>Figure 15 and 16</u>).

5. As a result of the higher level of "intelligence" the industrial robots will also be able to work in surroundings which are not optimally prepared for them. The development costs for problem-specific peripherals may thereby also be reduced.

6. Prices for industrial robots will not be influenced to a great extent in the foreseeable future. Despite more efficient and economic controls, the costs of devices will remain relatively stable, since machine building costs will not be lowered much.

7. There are already very many industrial robots, now they must be utilized!

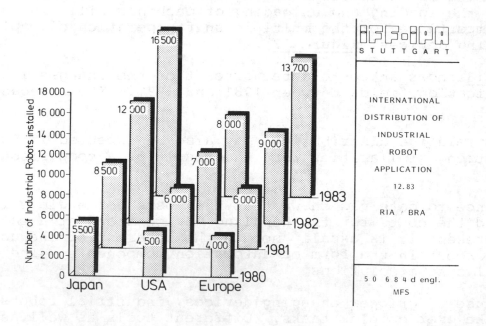

<u>Figure 1</u>: International distribution of industrial robot application (12.83).

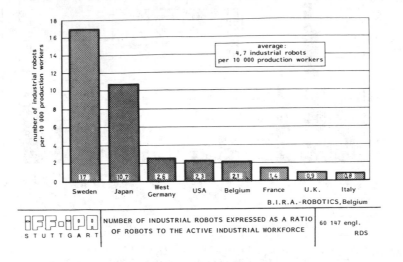

average:
4,7 industrial robots
per 10 000 production workers

| | | | | | | | |
B.I.R.A.-ROBOTICS, Belgium

iFF·iPA
STUTTGART | NUMBER OF INDUSTRIAL ROBOTS EXPRESSED AS A RATIO OF ROBOTS TO THE ACTIVE INDUSTRIAL WORKFORCE | 60 147 engl.

RDS

Figure 2:   Number of industrial robots expressed as a ratio of robots to the active industrial workforce.

	point-to-point control		continous-path control	
handling of work-pieces	loading and unloading of machines inter-linking	handling of workpieces and tools with com-hined gripper/ tool hol-der:	de-hurring polishing assembly	handling of workpieces and tools with com-hined grip per/tool hol-der (e.g. gripper and welding electrode)
handling of tools	spot-welding drilling	(e.g. grip-per and welding gun)	spray-painting de-hurring seam-welding assembly (e.g. handling a screwing me-chanism)	

iFF·iPA
unversität stuttgart | APPLICATION AREAS OF INDUSTRIAL ROBOTS | 56 271 engl.
RDS

Figure 3:   Application areas of industrial robots

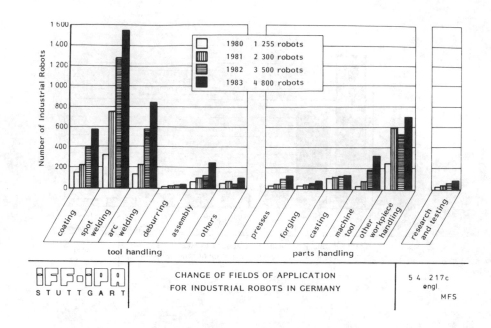

IFF-IPA STUTTGART | CHANGE OF FIELDS OF APPLICATION FOR INDUSTRIAL ROBOTS IN GERMANY | 5 4 217c engl. MFS

Figure 4: Change of fields of application for industrial robots in Germany

Figure 5: Undersealing of automobiles by an industrial robot (Daimler-Benz)

Figure 6: Industrial robot spotwelding side walls of
            automobiles (KUKA)

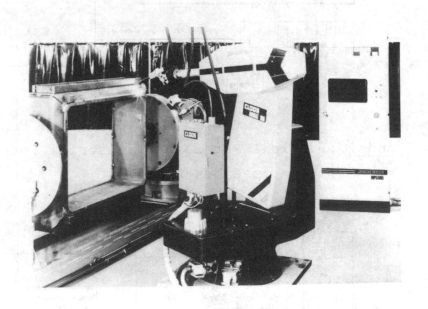

Figure 7: Welding work area with Romat 1o6 (Cloos)

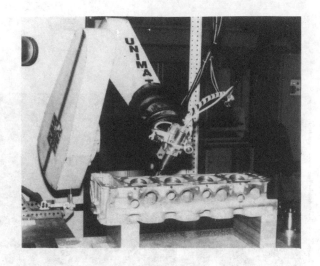

Figure 8: Industrial robot deburring a cylinder head.

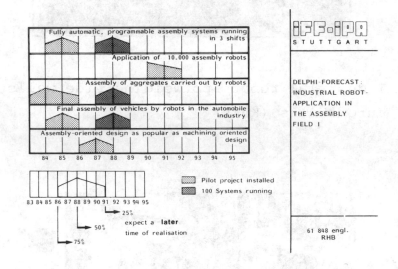

Figures 9 and 10: Delphi-Forecast: Industrial robot application in the assembly field.

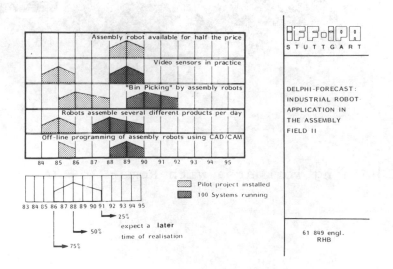

26

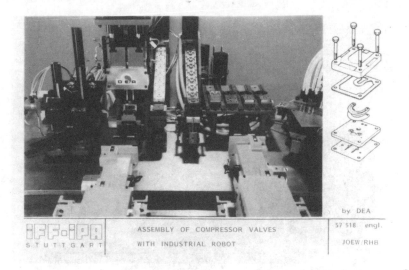

Figure 11:   Assembly of compressor valves with industrial robot.

Figure 12:   Loading and unloading machines with an industrial robot.

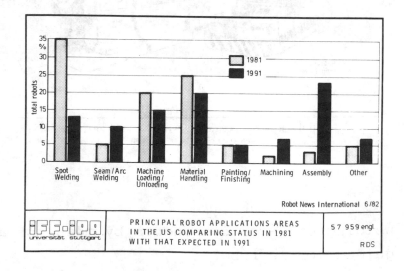

Figure 13:   Principal robot application areas in the US comparing
             status in 1981 with that expected in 1991.

27

Figure 14: Work place for the automated welding, grinding and handling of sheet parts

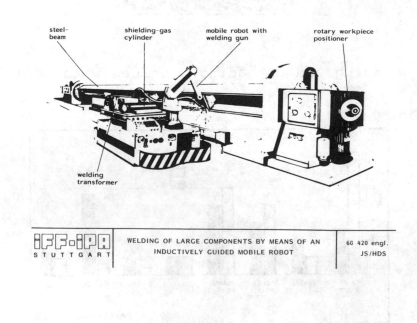

steel-beam

shielding-gas cylinder

mobile robot with welding gun

rotary workpiece positioner

welding transformer

Figure 15: Welding of large components by means of an inductively guided mobile robot

Figure 16: Test set-up of a mobile robot at the IPA.

Automation in Spray Finishing Systems

James R. O'Shea
The DeVilbiss Company

This paper will discuss automation as it is being
implemented in the world of spray finishing. It
will cover the various types of automated machines
used in current finishing systems, and also the criteria used in the selection of a particular machine
for a given application process. The basic principles
of electrostatics will be introduced and discussed as
they apply to both powder and liquid coatings. Appli-
cators used to spray both types of coatings will be
described and rotational atomizers for use with liquid
coatings will be highlighted. The need for continued
and more frequent use of automation in future systems
will be emphasized.

Automated finishing systems have today become the rule in many industries rather
than the exception. Many outside factors have influenced the practice of applying a
finish to a workpiece so that it may no longer be economically feasible to use manual
techniques. The factors influencing the movement to automate include:

a) the increased use of mass production techniques,
b) the increase in labor costs including the costs of fringe benefits,
c) the increased restrictions placed on manufacturing operations by regulatory
   agencies, such as the Environmental Protection Agency (E.P.A.) and the
   Occupation Safety and Health Act (O.S.H.A.) in the United States, and
d) the increased costs of petroleum-based products, including the solvents and
   paints used in the finishing industries.

Due to these factors, industry is moving toward automated finishing processes in an
attempt to hold the line on overall costs and also as a means of maintaining a consistency in the quality of their finishing products. This examination of automation in
the spray finishing industry will incorporate the discussions of several types of conveying systems, stationary as well as reciprocating spray guns, both liquid and dry

powder paints, electrostatic painting equipment, and even robotics.

On of the simplest machines used in automated spray finishing systems is the Chain-on-Edge machine. Some parts typically sprayed used a Chain-on-Edge machine include:

Telephone Bases	Telephone Receivers	Bottle Caps
Aerosol Cans	Toilet Seats	Insulator Caps
Brake Bands	Air Cleaners (Auto)	Auto Horns
Pails (Five Gallon)	Record Changer Turntables	Auto Wheels
Auto Arm Rests	Glass Bottles	Paper Cups
Bowling Pins	Beer Cans	Electric Motors
Electric Insulators	Flywheels	Frying Pans

Chain-on-Edge machines were developed to eliminate the handling of parts before they are dried or baked. The machine can be arranged so that the workpiece may be sprayed, then flash dried, and then conveyed through an oven for final baking. The workpiece may then be carried to another spray station for a second coating or to an unloading zone for transfer to another conveyor for use in assembly operations. One primary advantage of the Chain-on-Edge machine is that it offers great stability for small, fairly symmetrical work. The article to be sprayed is usually rotated in front of stationary spray guns at the spray stations. Although Chain-on-Edge machines can be designed to change elevations, ordinarily they are designed to carry the workpiece in the horizontal plane at one level. In practice, Chain-on-Edge machines are limited in both length and load by sliding friction. Depending on the weight of the workpiece, chain lengths up to 300 feet can be used with single drives. With multiple drives, however, it is possible to extend the chain lengths to almost 1,000 feet. In some systems, the spray guns used with Chain-on-Edge machines are continuously triggered. However, to obtain maximum efficiency, a skip spray, or control device, may be used to trigger the spray guns only when there is a piece to be coated in front of the gun. If the workpiece requires more coating than the spray gun can apply in the amount of time that it takes for the piece to move past the gun, a following device may be fitted to the machine at the spray station. This device allows the spray gun to be conveyed along with the workpiece for a short distance, so that the spray duration may be increased for that part. Once the spray cycle is complete, the follower mechanism is released and returns to its home position, ready to begin the spraying of the next part.

Large or irregularly shaped parts cannot be carried on a chain and, therefore, will usually be carried throughout a manufacturing plant on an overhead monorail conveyor. This is the most flexible type of conveying system, for it is able to accommodate changes in elevation. When parts are conveyed on monorail conveyors, spray stations may be set up using gangs of stationary spray guns carried on reciprocating spray machines, such as Long Stroke machines and Short Stroke machines. A stationary gun or a gang of stationary spray guns is the least expensive of the various methods of automated spraying systems. However, when the width of the piece being sprayed dictates the need for blending two or more patterns together to achieve the proper coating, there is the possibility of streaking with the use of stationary guns. Reciprocating spray machines may then be used to accommodate these wider parts, with the added benefit of reducing the number of spray guns required.

An alternative to using stationary spray guns with an overhead monorail conveyor would be the use of reciprocating spray machines. Long Stroke machines can be used to produce a uniform coating on flat work. Some of the parts typically sprayed using Long Stroke machines include:

Wallboard	Door Panels	Steel Sheets
Plywood	Table Tops	Carpet Backing
Acoustic Ceiling Tiles	Roof Decking	Stove Tops
Automobile Brakes	Asphalt Siding	Refrigerator Panels
Anti Scratch on Glass	Railrod Box Cars	Floor Coverings
Chalkboard	Mirror Backing	Mirror Silvering
Aluminum Sheets	Hides (Animal)	Automobile Floor Mats

The Long Stroke machine's stroke is usually 3 to 8 feet in length but may be extended to as much as 15 feet for special applications. The machines will coat work at the rate of about 120 square feet per minute for a first class finish and about 240 square feet per minute for finishes on surfaces such as ceiling tile and wall board. Both sides of large flat panels may be finished by a pair of staggered machines. The first machine will spray one side and the second machine will spray the opposite side of the work. Vertical Long Stroke machines are used with overhead conveyors. Horizontal Long Stroke machines can be used when parts are carried on flat or laydown conveyors. In the automotive industry where floor conveyors are used to carry the automotive bodies, vertical Long Stroke machines are used to coat the automobile sides and horizontal Long Stroke machines are used to coat the hood, roof, and rear deck surfaces.

When using Long Stroke machines, maximum efficiency is achieved if parts are uniform in length and conform to the machine spray stroke. Smaller parts may be grouped together on racks to conform to the spray stroke. The machines are equipped with cams so that the spray guns are turned off at the ends of the machine stroke. However, the guns will continue to spray between parts hung on a line or when an empty part hanger passes the machine. To eliminate this condition, a spray control may be provided to recognize gaps on the conveyor line and inhibit the spray guns from spraying during the absence of parts. If varying spray strokes are required, due to varying lengths of parts being carried past the machine, cams may be provided to allow the adjustment of the spray stroke. In some cases, encoders may be added to monitor the position of the spray guns within the machine stroke, so that external control systems may trigger the spray guns to accommodate more complex contours of parts as they pass the machine.

The guns mounted on these machines normally start spraying at the top or the bottom leading edge of the product. As the part advances, the guns move at a rate of speed (to a maximum of 288 feet per minute) to advance one spray pattern length in one complete machine cycle. The machine speed is adjustable to compensate for various conveyor speeds and work sizes. For higher conveyor speeds, additional guns are used to form a larger composite pattern to permit the relationship between spray pattern length and machine cycle to be maintained.

In addition to Long Stroke machines, Short Stroke machines may also be used to coat parts conveyed on an overhead conveyor. These machines are most often used with electrostatic spray equipment. The machine provides a simple harmonic motion, moving the spray gun through a short stroke of not more than 14 inches, and blends the patterns of the guns to provide a highly uniform coating on the product. The machine will hold up to 60 pounds, equivalent to the weight of eight spray guns. However, heavy duty models are available to hold as many as twelve guns. Electrostatic spraying with Short Stroke machines is ideally suited to automation since it recognizes area rather than the particular shape of a part. Therefore, different parts with like areas can be intermixed on the conveyor line while leaving the spray equipment on the same settings. It also has the added advantage of being more economical than other methods.

Electrostatic spraying is based on the principle that oppositely charged objects will be attracted to each other. Atomized paint particles are charged at the gun, then passed through an ionized field and are deposited on the grounded workpiece. A high voltage power supply is used to supply the charge at the gun by means of a high voltage cable. Overspray paint particles carried by air currents past the workpiece are attracted to the rear surfaces of the workpiece, providing the "wrap around" effect so often associated with electrostatic coating.

When using liquid coatings, atomization quality (atomized particle size) is important to electrostatic transfer efficiency. The smaller or lighter the atomized particle, the greater the influence the electrostatic field will have on attracting the atomized particle to the ware.

There are three types of paint atomization: Centrifugal or Rotational Atomization, Compressed Air or Conventional Airspray, and Hydraulic or Airless. Typical particle sizes achieved by these three methods are:

Rotational Atomization	.008 to .015 inches in diameter
Conventional Airspray	.012 to .015 inches in diameter
Airless	.015 to .025 inches in diameter

Rotational atomization is becoming more and more desireable in many automated systems for several reasons. First, it has a much higher transfer efficiency that other types of spray guns, reaching 90 percent efficiency in some applications.

Second, it is able to atomizer high solids materials, with up to 80 percent and higher solids content.

Rotational Atomizers use an air turbine to spin the atomizer bell at a high rate of speed, generally from 20,000 to 30,000 revolutions per minute. Paint is introduced to the atomizer from a paint line to a paint manifold behind the atomizer bell. The paint is directed to the rear cavity of the atomizer bell by a feed tube. The centrifugal force caused by the spinning or the bell forces the paint through small orifices in the wall separating the rear cavity of the bell from the outer atomizer surface of the bell. Once the paint reaches the outer surface of the atomizer bell, the centrifugal force breaks the paint up into small particles and throws the particles off the bell in a direction perpendicular to the axis of the turbine. Compressed air is supplied to a shaping air cap located on the front of the paint manifold and is used to contain the pattern of paint particles and to impart a forward velocity to the particles after they leave the edge of the atomizer bell. The high transfer efficiency of rotational atomization is due in part to the small size of the particles, the low velocity with which the particles pass the workpiece, and the electrostatic charge which is applied to the paint particles during atomization. Pattern size will usually range from 18 inches to 36 inches in diameter.

Two types of turbines may be used to provide the high speed rotation of the bell. An impulse turbine uses stationary nozzles which direct a stream of compressed air towards a bladed turbine wheel. The force of the air stream imparts a rotary motion to the wheel resulting in the high speed rotation. Unfortunately, the rapid cooling of the compressed air as it leaves the nozzles often causes any moisture contained in the compressed air to freeze as it contacts the blades of the turbine. This will sometimes cause the turbine to "seize up" or stop rotating and may reduce the life of the turbine. To prevent this condition, rotational atomizers using impulse type turbines will often require air heaters to increase the initial temperature of the compressed air so that after cooling, the final temperature of the compressed air is still above the temperature at which condensation will occur. Many times, a refrigerated air dryer will be used in an attempt to further reduce the possibility of turbine freezing. While the refrigerated air dryer will reduce the amount of moisture in the compressed air, an air heater is still recommended even with the use of the refrigerated air dryer.

The second type of turbine available is the reaction turbine. The reaction turbine incorporates nozzles which are a part of the turbine wheel itself. As compressed air is forced out of the nozzles, an opposite reaction imparts a rotary motion to the wheel, resulting in the high speed rotation. The reaction turbine is not susceptible to the freezing condition experienced with the impulse turbine, since the air streams are directed away from the turbine wheel. Neither an air heater nor a refrigerated air dryer is required when using a reaction turbine. Coalescing air filters and prefilters may be used to provide a clean and dry air supply for the turbine.

Turbine speed control is also an important consideration when selecting a Rotational Atomizer. Most Rotational Atomizers will slow down when paint is introduced to the atomizer bell. As a result, in order to achieve a rotational speed of 25,000 revolutions per minute, it is necessary to set the speed on the unloaded atomizer as high as 40,000 or even 60,000 revolutions per minute. The higher unloaded speed can accelerate bearing wear and shorten turbine life. To prevent this conditions, electronic devices utilizing both fiber optics to measure the rotational speed of the turbine and transducer control of the air supply to the turbine can be added as external controls to counteract the change in speed due to paint loading on the bell.

One Rotational Atomizer on the market includes a mechanical governor designed into the turbine which eliminates the need for any external device to compensate for changes in rotational speed due to paint loading. Utilizing a ball which seats on the orifice of an air valve, the governor compares the pressure of the compressed air used to turn the turbine with the centrifugal force created by a specific speed of rotation. A characteristic curve is established relating a specific air pressure to a specific rotational speed. When the speed is correct for the incoming air pressure, the centrifugal force will be just enough to cause the ball to seat on the orifice of the air valve. If, due to paint loading or for any reason in general, the speed of the turbine should decrease, the resulting decrease in centrifugal force will allow the ball to unseat, allowing more air through the orifice of the air valve. The resulting increase in air flow will increase the speed of the turbine until the speed is back at the setting where the centifugal force causes the ball to once again seat on the orifice of the air valve. This self-adjusting feature will allow no more than a 5 percent difference in speed between a loaded and unloaded condition, not enough to cause any noticeable effect on atomization quality.

Rotational Atomizers can be used either in a stationary arrangement or with Short Stroke reciprocating machines. When used in a stationary arrangement, the atomizers must be offset, usually by 24 inches diagonally. This placement insures that the patterns will not butt together excessively. Even with this offset, blending of patterns is still difficult since slight changes in flow rates or shaping air can alter uniformity. Short stroke machines offer an alternative to stationary mounting of the atomizers. The reciprocating motion, not more than a 14 inch stroke, tends to fill in the doughnut-shaped pattern of the atomizer. The paint particles appear as a soft cloud moving toward the grounded ware. The elimination of the doughnut-shaped pattern minimizes heavy edge buildup, since the pattern appears full. With the use of Short Stroke machines, atomizers can be stacked directly in line and the length of space parallel to conveyor travel required to obtain pattern bleeding can be dramatically reduced with superior uniformity.

While the majority of spray finishing systems are designed to apply liquid paint, dry powder paint may also be applied electrostatically and, depending upon the circumstances involved, may be more appropriate than liquid applications. Dry powder spray systems utilize paint supplied in powder form which is stored in a fluidized powder feed tank. The fluidization is achieved by introducing clean, dry, compressed air into the tank. Fluidizing the powder allows it to be pumped from the powder feed tank through a hose to a powder spray gun. The powder is electrostatically charged at the tip of the spray gun and directed toward the grounded workpiece. The powder sticks to the workpiece due to the electrostatic attraction. When the workpiece moves through a bake oven, the powder melts and flows evenly over the work, drying as a complete coating.

Other advantages of using powder application equipment are that high material utilization may be achieved because overspay collection systems can be designed to capture overspray powder and allow the overspray to be returned to the powder feed tank. In this way, utilization efficiency may approach 95 percent. Additionally, dry powder painting does not require the use of solvents, so emissions are of no concern. Also, the air used to collect the overspray may be filtered using high efficiency filtration techniques so that the filtered air may be discharged back into the spray room. These advantages must be balanced, however, against the cost of the equipment and the overspray collection systems, which may in some cases by higher than those in liquid spray systems, especially when multiple colors are being sprayed and colors must be automatically changed due to conveyor part loading conditions.

The preceding methods of automation represent the most common means of automating spray finishing lines. While less common, the use of robots provides the most flexible means of automation currently available, and for this reason, robots are becoming increasingly popular. Robots may be programmed to simulate the motions of experienced manual sprayers and will repeat these motions consistently all day, every day. Robots have been used to coat parts conveyed on all types of conveying lines as well as stationary parts. They are capable of carrying any type of spray device including electrostatic devices such as Rotational Atomizers.

Robots used in paint finishing systems are usually hydraulically driven devices utilizing cylinders and servo valves to achieve the motions required to manipulate the spray device. The robots may be mounted on transfer tables to increase the working envelope in order to follow parts moving on a conveyor line or to reach areas not normally accessible with the robot arm alone.

As the costs of raw materials required to provide a quality finish increase, including not only the cost of paint but also the cost of labor to apply the paint, automation will continue to find its way into the finishing industry. The automation needed to contain or reduce these costs ranges from simple systems such as Chain-on-Edge machines with stationary spray guns, to the more sophisticated systems including Long or Short Stroke machines with electrostatic spray guns, and even to the most sophisticated systems involving robots. The types of automation available in the spray finishing industry, the relatively low cost of the systems, and the ease with which automation may be incorporated into manufacturing processes make automation likely to become far more prevalent in the future.

References

Heath, D.A. "Criteria Used for Assessing Automated Coating Applications". (January
     1983)
Peter, D.A. "Automated Finishing Machinery". (January 1983)
White, W. H., Jr. "Automatic Coating Systems". (May 1982)
Unknown. "Automation". DeVilbiss Educational Services ES-88A (Date Unknown)

COMPUTER-AIDED DESIGN AND PLANNING

'APPLICATION OF CAD IN SINGAPORE'

L J YANG
Senior Lecturer
School of Mechanical & Production Engineering
Nanyang Technological Institute
Singapore

Abstract

Technology and industrial innovation are key elements to the growth of the economy. The introduction of computers, the use of NC/CNC machine tools and robots have greatly improved the flexibility and economics of batch manufacturing techniques.

Computer-aided design (CAD) is another recent innovation which has vast potentials for improving the productivity of the whole manufaturing system. CAD is also an important technology for product innovation and development.

Although the first CAD system was introduced about three years ago, to date, there are more than 203 workstations installed in 32 organizations in Singapore. The major users of CAD systems are

o   Architectural, structural design and mapping (54%)
o   Educational training (23%)
o   Precision Manufacturing Industries (14%)
o   Electronic and electrical (4%)
o   Shipbuilding and marine (4%)
o   Garment Manufacturing (1%)

CAD systems improve productivity, shorten lead times and allow modification and optimisation of the final products more effectively. The database established in the CAD system can also be used for production planning, tool and fixture design as well as generating NC tapes for machine tools. Not only will these be done in a small fraction of the time previously required, but there will be no dimensional discrepancies with the data originally specified by the designer.

Since Singapore has a good industrial climate, the CAD system will definitely enhance the growth of the economy.

1. INTRODUCTION

Technology and industrial innovation are important to the economic, environmental and social well-being of the people in a country. They offer an improved standard of living, increased productivity of public and private sector, creation of new industries and enhanced competitiveness of products in the world markets.

Although some of the pressure for innovative action in industry comes directly from external stimuli, the majority of actual innovations arise from within the product design or manufacturing areas. In some cases, the stimulus arises from the desire to improve product performance or quality or to increase profitability by reducing production costs.

The introduction of solid-state digital devices and computers have contributed greatly to the advances in manufacturing technology. The use of NC/CNC machine tools has made batch production of quality component parts more economical. The use of machining centres with automatic tool changers, automated inspection devices and robots have further improved the productivity and flexibility of manufacturing systems available nowadays.

Computer-aided design (CAD) is the most recent innovation which will benefit greatly the future of our manufacturing industries. With a CAD system, an engineer can work interactively with a computer, using graphic displays to develop and record his design. Drawings and parts lists are produced by computer driven drafting machines and printers.

CAD is also the key to the future development of an automated factory. It defines the data-base that is internal to a manufacturing company at the design phase of a product. CAD systems enable this data to be reduced to the basic binary blocks which are the currency of electronic systems needed for production and inventory control procedures. The database also includes the geometric shape information needed to program NC/CNC machines, or to design holding fixtures, moulding dies, press tools etc.

The development of modern software will also enable manufacturing engineers to access this database and generate the appropriate manufacturing control tapes and specify all the tools in detail. Not only will this be done in a small fraction of the time previously required, but there will be no dimensional discrepancies with the data originally specified by the designer.

CAD systems allow designers to modify and optimise the final products more effectively. The more accurate CAD drawings also reduced the need for, and costs of, manufacturing prototypes. CAD is therefore an important technology for product innovation and development. It is also the most essential tool we cannot afford not to have to improve productivity and product quality. This is especially so when we are facing competitions from the developed countries where fully automatic plants have been set up and the less developed countries where the wages are very much lower (1).

2. STATUS OF CAD APPLICATIONS IN SINGAPORE

Although the first CAD system was introduced around 1981, to date there are more than 203 workstations installed in 32 organizations in Singapore.

The major systems are supplied by Computervision, Intergraph, Calma, Gerber and IBM.

The applications of the CAD systems are mainly for

o   Architectural, structural design and mapping.
o   Educational training.
o   Precision manufacturing industries.

o   Ship-building and marine applications.
o   Fabric Pattern layout

Figure 1 shows the major CAD users in Singapore.

(a) Architectural, structural design and mapping (54%)

This is the biggest area where CAD is used in Singapore. There are altogether 109 workstations installed in 13 organizations. The biggest single user is the Housing and Development Board (HDB) which alone has 40 workstations which are using the Computervision system. The Public Works Department (PWD) has 19 workstations which are using the Intergraph system. Other users are the Public Utilities Board, Ministry of Environment, URA and a number of architectural firms.

The main applications are:

o   Architectural and structural design of buildings, roads.
o   Preparation of detail drawings.
o   Mechanical and Electrical services
o   Mapping

(b) Educational training (23%)

There are altogether 47 CAD workstations installed in Nanyang Technological Institute, National University of Singapore and the EDB training centres.

The three Schools of Mechanical and Production Engineering, Electrical and Electronic Engineering, and Civil and Structural Engineering of Nanyang Technological Institute have a total of 24 workstations running on computervision, VAX 11/750, Prime 750 and ICL PERQ computers.

The students from the School of Civil and Structural Engineering will be trained on structural design while those in the School of Electrical and Electronic Engineering on printed circuit board (PCB) and integrated circuit (IC) design.

During the 83/84 academic year, all the second year students from the School of Mechanical and Production Engineering received their basic in-house training on CAD based on the HP 9836 computer with CAD 200 software and the Tetronix 4054 computer with Radraft software from Radan. Each group of students spent 2 days on two workstations, familiarising themselves with the basic commands as well as preparation of engineering drawings.

The final year students from the School of Mechanical and Production Engineering will receive their training in CAD in the following areas:

o   Optimisation and Drafting using VAX 11/750 and OPTIVAR and DOGS software.
o   FEM modeling and Analysis using VAX 11/750 and PAFEC software.
o   Turnkey Applications.
    -   Plastic Mould Design, Plant and HVAC design using Computervision system with CADDS, CVMD and CVPD softwares.

The optimisation and drafting project involves the use of non-linear optimisation technique to determine the optimum solution for a mechanical design. The students will learn how to formulate the problem, select the equality and inequality constraints, utility functions and determine the trade-offs for the various parameters. Students will then write a parametric design program to automate the drafting process.

In the finite element modeling and analysis project, students will use the PAFEC software for the finite element analysis of a plate with a hole under uniform pressure loading. The students have to select a finite element model and comment on the choice of the mesh; validity and accuracy of the computed results. They will devise a systematic approach to determine the size and location of holes to reduce stress concentration and make a refine finite element model.

In the applications of turnkey system, the students will choose an application software in the Computervision System that is relevant to their option of study. As such the Production students will be required to use the basic mechanical design software to design a plastic mould for a given product and prepare the NC tape for computer aided manufacturing. The Thermo-fluid students will be required to design the air conditioning system and piping layout for a given building according to the standard code of practice.

The Computervision EDB CAD/CAM Training Unit (CECTU) was set up jointly by Computervision Corporation of USA and the Economic Development Board in February 1983. it has 15 workstations and provides training in CAD/CAM in the area of:

o   Mechanical design and drafting
o   Numerical control
o   Structural engineering design
o   Printed circuit Board design
o   Intergrated circuit design.

To date, more than 300 people have been trained in various areas of CAD/CAM application.

(c) Precision Manufacturing Industries (14%)

There are altogether twenty nine (29) workstations installed in four major organizations. One company has alone 16 workstations which are using the Computervision system.

The CAD systems are mainly used for the design and drafting of precision mechanical component parts. They are also used for the design of fixtures, toolings, preparation of operation sequence sheets as well as NC tapes for machining.

(d) Electronic and Electrical Manufacturing Industries (4%)

There are altogether eight (8) workstations installed in four major organizations. They are used for printed circuit board design as well as other applications. One of the semiconductor company has purchased a Calma computer-aided integrated circuit design and manufacturing system. This will be connected to other computers in the company's design centres elsewhere.

(e) Shipbuilding and marine applications (4%)

There are eight (8) workstations with two major organizations. They are used for structural, mechanical and NC applications.

(f) Garment manufacturing (1%)

There are two (2) workstations with a garment manufacturer which uses the CAD system for fabric pattern layout.

3. THE BENEFITS OF USING CAD

The benefits arising from the use of CAD are:

(a) Higher productivity on drafting due to
  - higher operating speed
  - less time spent on modifications
  - parts can be recalled from the memory of the system.

40

(b) Essential tool in designing certain products like integrated circuits.

(c) Effective tool for modifying and optimising the final design of a product.

(d) Shorter lead times from designing to manufacturing.

(e) Better organisation and numbering of component parts.

(f) Same database can be used for the design, production planning and preparation of NC tapes.

(g) Reduction in material utilization due to optimisation of design and nesting.

(h) Reducing the need for, and costs of, manufacturing prototypes.

(i) Capable of checking interference prior to plant construction and piping layout.

(j) Better control of proprietary information.

(k) A useful marketing tool.

To sum up, depending on the actual application, CAD is beneficial in at least five areas, namely

o   In product design, without CAD, the design would either be impossible or sub-optimum.

o   In process, CAD is an efficient technique.

o   In lead-time, without CAD, product would reach the final market place substantially later.

o   In marketing, CAD allows a product design to be optimised in a 'real time' interaction between the customer and the supplier.

o   In control, CAD allows the management more effective utilization of design data and proprietary information.

4. CONCLUDING REMARKS

The above section clearly indicates that CAD is an important technology for product innovation and development which is the key element for enhancing the growth of the economy.

Gee (2) has studied the various factors contributed to the success of both the post war West German and Japanese economy. These are:

o   An industrious and frugal people.

o   A labour-management relationship founded on consensus and participative management.

o   A framework for close government-industry cooperation.

o   A free-enterprise commitment where weak, non-competitive firms are allowed to go out of business.

o   A heavy industrial R & D emphasis self-financed by industry in the main.

o   An unimpeded, bidirectional flow of technology across national borders.

o   The existence of linker organizations to expedite the transfer of technical information and technology.

o   A broad system of fiscal and regulatory incentive to stimulate all stages of the innovation process and

o   Special attention to small and medium-sized business enterprises.

Since the industrial climate of Singapore is good, CAD will definitely speed up our development on product design and manufacturing processes; improve the productivity and quality of our products; and consequently enhance the growth of our economy.

ACKNOWLEDGEMENT

The author would like to thank Miss Aminah for typing the manuscript.

REFERENCES

(1) Business Week, March 28 1983. pp. 65-67.

(2) Sherman Gee - "Technology Transfer, Innovation, and International Competitiveness", John Wiley & Sons, 1981, pp. 159.

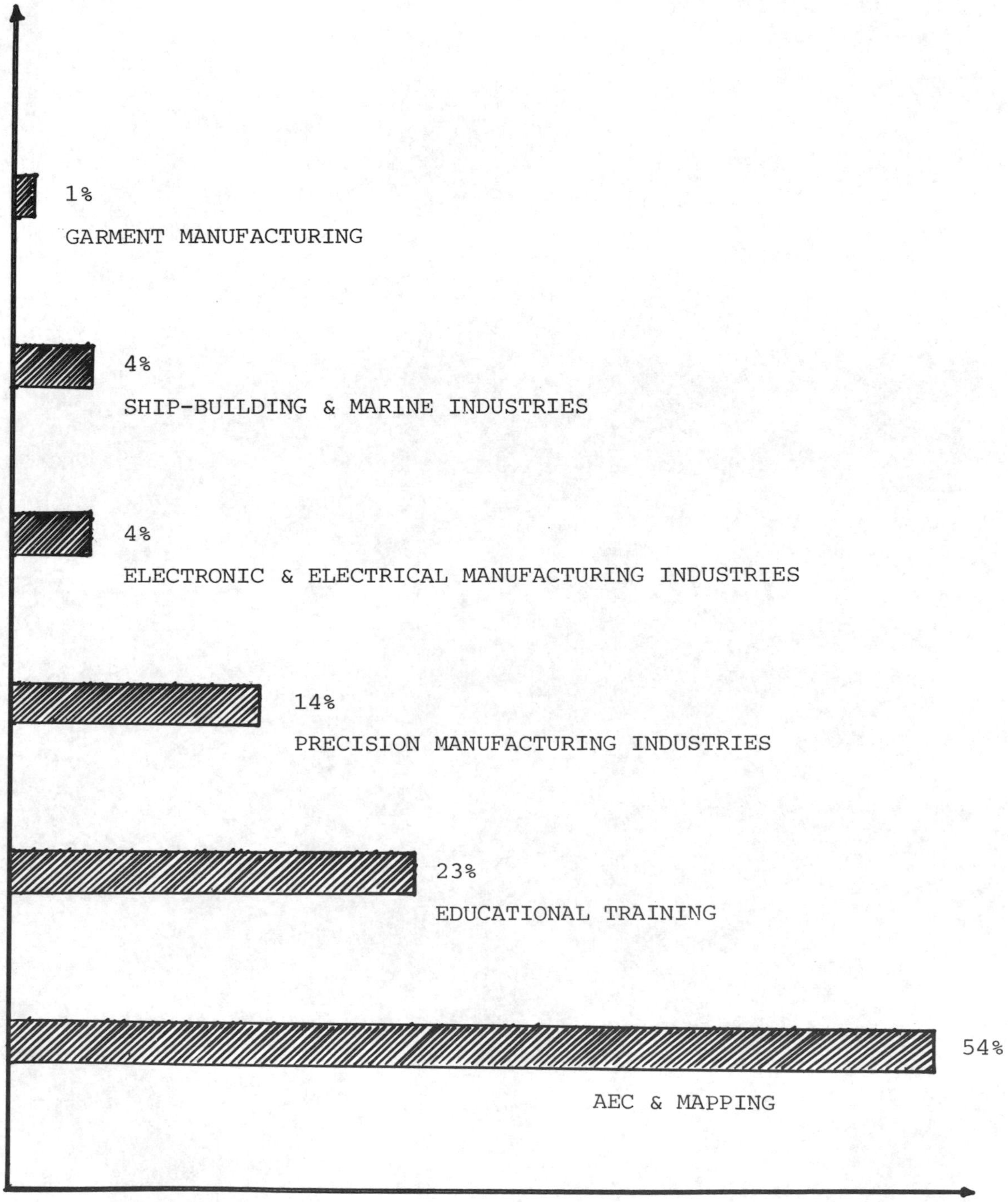

1%
GARMENT MANUFACTURING

4%
SHIP-BUILDING & MARINE INDUSTRIES

4%
ELECTRONIC & ELECTRICAL MANUFACTURING INDUSTRIES

14%
PRECISION MANUFACTURING INDUSTRIES

23%
EDUCATIONAL TRAINING

54%
AEC & MAPPING

FIGURE 1   THE MAJOR CAD USERS IN SINGAPORE

43

SOME ASPECTS OF CAD IN DESIGN OF MECHANICAL COMPONENTS AND SYSTEM

A.K.Abd El Latif   and   T.T.El-Midany

Professor   Asst. Professor

Dept. of Prod. Eng. and Mech. System Design
College of Engineering,King Abdulaziz Univ.
Jeddah,Saudi Arabia

ABSTRACT

Industry recognize now the vital role that CAD will play in our future and both lay stress on the need to invest more and to expand the existing CAD systems. The trend is to integrate design process, data, analysis , optimization and modeling, in a manner which eliminates a great many of the manual steps of more conventional process. CAD is a multi-discipline approach, offers the advantage of integrated team-machine work. The interactive CAD systems have been given wide recognition as being at the fore front of CAD in the mechanical technology. It is moving very quickly on both the software and hardware fronts which make it more suitable for extension without interruption onto several areas of mechanical technology. In the following we present a comprehensive views for all the important aspects of CAD techniques in the mechanical design. In this context, not only from the point of view of developments and application of hard and soft ware of CAD systems but also on the logics of the design process.

INTRODUCTION

Computer have been in use for CAD since the early 1960's within manufacturing companies, initially as a large expensive main frame, and more recently in the form of dedicated mini-computers operating in a time-sharing mode with several graphic and alpha-numeric screens carrying out engineering design tasks. Computer aided design can be described as any design activities that involves the effective use of the computer to operate or modify an engineering design. The increasing use of CAD systems has helped the drawing office to improve time scales and to produce drawings to a consistently high quality. Recent developments make it profitable to bring CAD to the working place of an engineer.

Developments of the computer Aided Design, CAD Techniques for Mechanical design in the last ten years give a good chance to realize that CAD is really economic instrument for the mechanical industry. Industry recognize now the vital role that CAD will play in our future and both lay stress on the need to invest and to expand more in it.

Computer graphics and, indeed, the more wide spread application of computing in the field of design engineering really began to have an impact in the years 1963-1965. Then after an initial learning period it was predicted that great things would happen with CAD due to developments in graphic terminals, macrostorage technology and the reducing cost of central processor power. However limitations imposed by the high cost of running the systems relative to traditional methods, which is itself was a manifestation of the fact that hardware and software limitations restricted performance, [1].

CAD is an integration of all phases of the design process through computer-modeling of the design components or systems. The building of data structures that describe the models in a way that is amenable to subsequent analysis, modification. Simulation, and graphic representation,is the main feature of computer aided design. It is the synergy of these computerized phases of design, and not the computerization itself, which warrants a new name for this activity.

CAD represents an excellent tool by which we can integrate the engineering design process, data, analysis, optimization and modeling, in a manner which eliminates a great many of the manual steps of more conventional process. The aim of CAD is to study and implement computer applications in the process of design-development-manufacturing. The computer facilities is used to help the designer work better, faster and more efficiently. The levels of use of the computer vary according to the application field and the firm's willingness to use these techniques.

The design process itself differes widely from industry to industry and the field of effective computer application will in each case have to be determined pragmatically CAD is not yet a unified and standardized technology and its characteristics vary depending on the kind of industry, factory, location, whether it was installed locally or for special purposes or bought ready made, what makes CAD technology fundamentally differ from other is that it came through comprehensive study of a virtual object. The product is-or should be-concieved, adjusted, tested without any actual model or prototype being built before the final step. A considerable amount of time is thus clearly saved. Accordingly, many studies instead of one could be carried on at a lower cost. Hence quality, reliability are increased and manufacturing costs are reduced. The use of CAD in the last two decades showed many advantages including faster design, cheaper design, better design automated draughting, less labour, and the most important being the saving of time, thus reduce investment and operation costs. In this way CAD become increasingly important and a great effort is directed to produce design packages and efficient algorithms for the mechanical design of components, units and systems in industry, [2].

There are many benefits of CAD,only some of which can be easily measured Some of the benefits are intangible, reflected in improved work quality, more pertinent and usable information, and improved control, all of which are difficult to quantify, other benefits are tangible, but the saving from them show up for down stream in the production process, so that it is difficult to assign a dollar figure to them in the design phase. A checklist of potential benefits from implementing CAD as part of integrated CAD/CAM system as follows:

1. Improved engineering productivity
2. Shorter lead times
3. Reduced engineering personnel requirements
4. Customer modifications are easier to make
5. Faster response to requests for quotations
6. Avoidance of subcontracting to meet schedules
7. Minimized transcription errors
8. Improved accuracy of design
9. In analysis, easier recognition of component interactions
10. Provides better functional analysis to reduce prototype testing
11. Assistance in preparation of documentation
12. Designs have more standardization
13. Better designs provided
14. Improved productivity in tool design
15. Better knowledge of costs provided
16. Reduced training time for routine drafting tasks and NC part programming
17. Fewer errors in NC part programming
18. Provides the potential for using more existing parts and tooling
19. Helps ensure designs are appropriate to existing manufacturing techniques
20. Saves materials and machining time by optimization algorithms
21. Provides operational results on the status of work in progress
22. Makes the management of design personnel on projects more effective
23. Assistance in inspection of complicated parts
24. Better communication interfaces and greater understanding among engineers, designers, drafters, management, and different project groups

Most aerospace companies have moved into computer-aided design via the implementation of interactive drafting along a broad front. The interactive computer aided design system has been given a wide recognition both inside and outside the aerospace industry as being at the fore front of CAD Technology. They developed P & WA interactive design process which differ from the approach being employed in most other companies. The application is complete in the sense that it extends from preliminary design through fabrication of a part; it is modular in the sense that it can be used on one component or on several; it is heavily design-analysis oriented rather than design/drafting.

The objective in each instance has been to start with preliminary design, to have the part description and properties gradually evolve,and to terminate the design phase of the process with a complete numerical description of the part, including all tolerances. Associated aerodynamics, heat transfer, stress and vibrational analyses in conjunction with the analysis of part life, durability and other properties are carried out as the design progress.

The description of the part of components, together with the associated analytical properties, is built up in the integrated data base. This data base serves as the principal communication medium among engineering groups participating in the design, Fig.1. The final part description is made available to manufacturing to continue the interactive process-in process planning, tool design, N/C programming. [3].

A real benefit of using a CAD system is that after the part's drawing is created, it can be stored in the computer's memory (or on disk or tape) and instantly retrieved later for duplication or further modification. Once the part is in the computer, drawings can be created by using automatic drafting machines. It can produces extremely accurate prints in minutes instead of hours or days, [4].

In most CAD systems, a wide variety of analytical software is available for mathematical testing and modeling of part designs. For instance programs can automatically calculate the volume, surface area, and weight of a part. CAD users can establish finite element models.

Once the part's geometry is defined, its geometry, then becoms the basis for future part machining operations and indeed, even the tool and fixture design for the equipment needed to manufacture the product. For instance, for a part that needs turning on a lathe, the CAD software defines the cutter location (CL) or tool path , for the part's geometry. This CL becomes the basis for further computer processing to turn it into numerical control (NC) instructions to run specific machine tools,[5]. Our experience to date indicates that the mechanical community is appreciably lagging their electrical/electronic counterparts in the use of computer assist devices to aid in the design and building of their products.

EFFICIENT DESIGN ALGORITHMS

At the risk of being some what controversial, we find it difficult to make a clear definition of "design engineering." Many people will feel that it is bound up with product design the eventually aim of which is to produce drawings of components or system that will enable a product to be manufactured. This may be too narrow a view.

Hence there is no basic theory or fixed definition for design. However some ideas consider design as a mapping of a point in the attribute space. Normaly, the identified need specifies the functions a new product or system has to perform. The functions are used as a starting point to arrive, through a design process, at a physical description, or attribute, of the final product or system. Design is regarded as the set of activities leading from the establishment of a product requirement to the generation of the information necessary for making product, under real conditions the designer has to use a strategy to find the mapping points by giving through a series of steps. Thus design is a sequence of steps which aim at turning an abstract idea into a concrete system or structure. These steps may be placed in the form of morphology chart as shown in Fig.2 [6]. The various design-related tasks which are performed by a modern computer-aided design system can be grouped into form functional areas: Geometric modeling, engineering analysis, design review and evaluation and automated drafting.

The resulting morphology, then, placed in an orderly fashion the sequence of decisions which would be adequately resolved in order to emerge with an effective set of plans for the needs which have been identified. After many expensive, abortive trials in the development of complex systems and designs among a diverse array of disciplines and among people of different training and abilities. The structure or form of which provides the frame work for the decision necessary to accomplish system and design development were developed by Asimow, design morphology In attempting to plan for the most efficient use of resources the Asimow phases of design are adapted for design Fig.3.

From the recognition of needs in the feasibility study to the last step in retiring the system a great deal of knowledge is gained concerning the design being developed. Hence the iterative nature of design is recognized as an integral part of the process i.e. where new factors are involved in an engineered design, then iteration is both normal and desirable throughout the design process. Similar set of activities and new literature on systematic design methods appeared in the last ten years however it stem from the Asimow design morphology Fig.3.

Within the design process there is a contradiction at the level of the human/machine intraction itself. The human being may be viewed as the dialectrical opposite of the machine in that he or she is show, inconsistent, unreliable but highly creative. The machine on the other hand may be regarded as fast, consistant, reliabled but totally non creative. Initially these opposite characteristics were perceived as complementary and regarded as providing the basis for human/machine symbiosis. Such a symbiosis would however imply dividing the design activity into it's quantitative and qualitative elements. That is to say into its creative and non creative. The notion then is that the non creative elements may be allocated to the machine and then creative elements left to the human beings. This is a Taylorist notion and implies at the level of design the equivalent of separating hand and brain within the field of skilled manual work [8].

The design activity cannot be separated in this arbitrary way into two disconnected elements which can then be added and combined Fig.4. The process by which these two opposites are united by the design to produce a new whole is a complex, and as yet it researched area. The sequential basis on which the elements interact is of extreme importance. The nature of that interaction and indeed the ratio of the quantitative to the qualitative depends on the commodity under design considerations. The very suitable process by which designers review the quantitative information they have assembled and then make the qualitative judgement is extremely complex, and much freedom must be left to the design in doing it.

A great percentage of the design work consists of seeking of data and standards, or looking up catalogues and drawings in other words his first step is acquiring and grouping data. This is a time consuming part of his work which could be replaced by storage packages, file organization and retrieval programs. Another typical part of this activity involves calculations, most of which are generally used long series of routine calculations. These can be computerized in the form of series of independent subroutines, and so a well selected subroutines library and a combined retrieval system may prove useful to the designer.

The real creative part of the designer cannot be automated because intution is a quality reserved for the design, but it can be helped greatly by the computer. The whole CAD work is a mosaic of deterministic routine works, indeterministic decisions, and creative parts. The realistic solution is a system, or a language, characterized by a high interactivity. Some, points of the programs can be equally well handled in interactive or deterministic ways. The interactive version tends to be shorter and simpler because the deterministic type of program must contain all possible versions, decisions and essentially longer diagnostics, than a program written in interactive style. [9].

COMPUTING TECHNIQUES AND SYSTEMS

Hardware

Computers are generally considered to fall into three size categories, all of which are based on roughly the same architecture as that shown in Fig.5. The three size categories are:

1. The large main frame computer
2. The minicomputer
3. The microcomputer

The large main frame computer is distinguished by its cost, capacity,and function. The main memory capacity is several orders of magnitude larger than the minicomputer, and the speed with which computations can be made is several times the speed of a minicomputer or microcomputer.

Minicomputers are smaller versions of the large mainframe computers. The trend toward miniaturization in computer technology provides two alternative approaches in the design of a computer. The first is to package greater computational power into the same physical size with each new computer generation. The second approach is to package the same computational power into a smaller size. The minicomputer manufactures elected the second approach in developing their product lines. Smaller minicomputers often overlap with microcomputers in terms of the functions which they perform. These microcomputer/mini computer functions are different from those performed by large   frame. Tables (1) and (2).

Table 1. Two typical applications performed on large mainframe computers and large minicomputers

---

1. Complex Engineering and Scientific Problems

    Examples include iterative calculation procedures often required in heat-transfer analysis, fluid dynamics analysis, or structural design analysis. These calculations are typical of computer-aided design applications.

2. Large-Scale Data Processing

    Examples include corporate accounting and payrool operations, production scheduling, compiling production costs, and maintenance of large information files. Some of these examples are found in indirect computer-aided manufacturing applications.

Table 2. Typical characteristics of microcomputer minicomputer applications.

---

1. The computer is a system component. The overall system, which might be a piece of test equipment, a machine tool, or a banking terminal uses the small computer much as it might a switch, power supply,or display. The computer may not even be visible from the outside.

2. The computer performs a specific task for a single system. It is not shared by different users as a large computer is. Instead, the small computer is part of a particular unit, such as a medical instrument, typesetter, or factory machine.

3. The computer has a fixed program that is rarely changed. Unlike a large computer, which may solve a variety of business and engineering problems, many small computers perform a single set of tasks, such as monitoring a security system, producing graphic displays,or bending sheets of metal. Programs are often stored in a permanent medium or read-only memory.

4. The computer often performs real-time tasks in which it must obtain the answers at a particular time to satisfy system needs. Such applications include machine tools that must turn the cutter at the right time to obtain the correct pattern,or in missile guidance where the computer must apply thrust at the proper time to achieve the desired trajectory.

5. The computer performs control tasks rather than arithmetic or data processing. Its primary function might be managing a warehouse, controlling a transit system, or monitoring the condition of a patient

---

Computers have been in use for CAD since the early 1960's within manufacturing companies, initially as a large expensive main frame, and more recently in the form of dedicated mini-computers operating in a time-sharing made with several graphic and alpha-numeric screens carrying out engineering design tasks. Recognising the need to cater for industrial users of mini-computers in CAD and with CAM environment, most of the major mini-computer suppliers have now introduced a more powerful range of machines known as super-minis. There have a large addressing capability and can cater for much larger engineering programs in their main memory,

Fig. 6-a .

Recent developments in computing hardware manufacture indicate that due to decreasing physical size, increasing performance and reduction in cost, a situation will be created to provide local computing power within each individual workstation. These individual computing workstations will have the capability to store and address large engineering programs for both design and manufacture, with interfacing and communication facilities to a central host computer. Large disc storage, printing and high speed plotting facilities will be available from the host computer Fig. 6-b. Systems of this type are known as distributed systems, and further developments indicate that links between several computers and workstations will be introduced providing a networking capability to improve still further the efficiency of the system Fig.7, [12].

Computers are generally described by their word length, i.e.8 bit 16bit 32 bit .. et. which represents the width of the internal data transfer paths of the computer. The longer the word length the larger and more complex the CAD system process that can be run. A typical categorization used to be microcomputers (8 bits), minicomputers (16 bits) and main frames (32 bits) or more. However longer word length micro and mini-midi has been in production for several years. Historically CAD systems were implemented on main frames because only such computers were able to support the high computational loads imposed by these large interactive systems.

There are two basic methods of organizing the computer hardware, as a computeralized system or as a distributed system. The centralized system would be based on one of the larger mainframe capable of supporting twenty or more simultaneous users. Problem imposed are, if the main frame goes down all work will cease, and if a company has two or more well separated sites. Here the solution would be to place remote terminals at the satellite sites connected to the central main frame by high speed, wide band land lines or by a microwave link. This will be both expensive and questionablly effective. The distributed system comprise two or more computers, micros, minis, super minis or even main frames linked together. They can be sited with the user population and so provide dedicated support. If one of the computers goes down the only the one site will be effected.

Developments in micros and minis are still proceeding at such a pace that it is very difficult to identify the type of equipment that would be the best investment. Looking only slightly ahead, it would seem that we should be thinking in terms of an arrangement something like that shown in Fig.8. In this, the development of a hierarchy is envisaged in which at the user level we have clusters comprising, say, five VDUs and the graphics terminal although it may be unnecessary to make the distinction in the future. The VDUs would be "intelligent" terminals capable of being used for small computations and sharing a local printer. Two or more clusters would be linked to a mini-machine which can handle jobs requiring, say, 30K in time-shared mode and will also service other shared facilities such as the local data-bank and graph-plotter. Program files would be stored at this level but it may be more economic to store user's files as personal discs. The mini-machine in turn would be linked to regional resources and hence give users an opportunity to initiate large jobs (such as FLOWPACK) from local terminals. Users would also gain access to the larger data banks this way. [10].

Software

Without the tremendous advances made by hardware suppliers most of today's software products would not be able to perform as they do software is however the ingredient that turns these rather expensive boxes into these useful tools for a company and it is portable software that makes the general purpose mini computer so cost effective against the turnkey systems.

Historically, software has been a low cost item, often bundled', that is, sold together with the hardware at a fraction of the cost of production (to ensure the hardware sale), such a production route also infers a number of limitations.

Software development was performed with an in house development team, and there was little room for the soft specialist with engineering background external to the company. Regrettably, however, this is becoming increasingly expensive, and this has provided the stimulus for the emergence of specialist technological software vendors with general purpose software products. [13].

General purpose software may be viewed in the same technical and economical light as hard-ware components in manufacturing. The design and development costs are spread over a great many users, thus rendering it uneconomic to expend valuable brain power on this particular piece of hardware. The analogy with software is clear. The marketing, stockholding and servicing costs are also spread over a great many users, whilst the full technical resources of the full technical resources of the supplying company are also available to solve both technical problems in the use of the system and implementation of the system on the uses machine.

Many software appeared which have general use, SAMMIE has been widely used for design studies in a number of fields from the design consultancy point of view , for instance, leyland vehicles have used it to evaluate the cab design and particularly front and rear visibility of a heavy truck. Moving along the design we come to DRAGON, this is essentially a two dimensional package which was originally developed by George Wimpey MEXC. It's particular strengths are in schematic layouts and detailing, and it has aroused considerable interest as a low cost first step in CAE for many companies. MOSAIC introduces a new concept to the analysis of component subject to heat and mechanical loading using the finite element technique. The engineers with little or no previous experience can apply MOSAIC to model his components.

Our experience to date indicate that the mechanical community is appreciably lagging their electrical/electronic counterparts in the use of Computer assist devices to aid in the design and building of their products.

GEOS has been actively pursuing applications for interactive graphics since 1974. The introduction of three-diemensional software by graphic vendors has made possible the construction of complex geometric data in a manner competitive with conventional manual methods. The GOES system CAD/CAM system for producing mechanical parts is relatively simple in its concept yet its simplicity is an important factor in its success. Mechanical parts are designed on an interactive graphic machine. Once the design is approved, the detail parts are extracted, dimensional, and prepared for transmission to a plotter for hard copy record.

The overall data base has been structured within the standard capabilities offered by the graphic machine. These standards serve a number of purposes, mainly to insure that tool designers and NC programmers have easy access to the geometric data they need, without being burdened with data which is of no value to their particular function. The advent of 3D software has made the generation of complex mechanical designs on interactive graphic systems much more practical than was the case prior to 3D [12].

The analyzing of mechanical design is very important and usually occurs at a critical time in the design cycle, namely after the conceptual designs have been created and production designers and detailers are ready to move into the cycle overdesigning of the parts can be very costly from both a material and labor standpoint. Under-designing can be disastrous from a functional and safety standpoint. Analysis which can be performed quickly and accurately is a major benefit. The analysis software, which is a subset of the graphics 3D operating system, provides an effective tool for the construction, modification, and viewing of finite element models. See Fig.9 for typical finite element model and for typical thermal analysis modelling.

The creation of mechanical designs via IAG establishes a design data base from which piece part definitions can be easily extracted. Soft techniques such as mirroring rotating, copying, and use of component libraries help to insure cost effective defination of the piece parts. [13].

The tool designer utilizes the system in the same manner as the mechanical parts designer, with the added advantage of having the parts geometry visible, around which he can construct his tool.

Comonly Used Systems

EUCLIP-The True CADD

In 1970, EUCLIP was conceived from the need to model complex 3 dimensional solid forms, for the purpose of performing a design analysis. In 1980, EUCLIP was commercially introduced as the only truly 3 dimensional solids-based CAD system, i.e. working directly on the object to be designed (mechanical parts, etc.). Since the beginning, the developed team [14], has always remained consistent with its design philosophy, See Fig. 10. The six principles of this philosophy which makes EUCLID unequalled are: Integrated approach, flexibility, portability, solid modeling, coherence and ease of use.

A true CAD system must cover the total requirements of the design and manufacturing organization namely: design, analysis, simulation, documentation and drafting, manufacturing engineering, storage. Such a system must not be limited to the simple function of an 'electronic drawing board'. Only a 3 dimensional solid modeling approach meets all these requirements, including in particular the essential need for an interaction between form and analysis.

The design process varies with the application, the specific company and with time. Thus a CAD system must be open, i.e. capable of adapting to the existing situation (procedures, analysis programs already used, etc.) and capable of evolving with the company's development. The technological evolution of data processing hardware is occuring faster than the development of software systems. Therefore, it is essential that the CAD system be really "portable", i.e. independent of the hardware. The objects to be designed are 3 dimensional by nature whereas the traditional tools used in design (paper, drawing board...) are necessarily 2 dimensional. This contradiction generated the traditional industrial drawing and descriptive geometry techniques. Drawings provide excellent information transfer, however they are not good for conceptualization as the designer must "think 3D". When a designer wants to design an object he instinctively visualizes that object, not a front and a top view. With EUCLID these 2 dimensional constraints can be removed and a true 3 dimensional solid model of the object can be built with all its proper geometric characteristics (and with an unequalled degree of accuracy).

It is then possible to display this virtual model, to move, it to disassemble or assemble its elements, to simulate its possible movements, as could be done with the real object. The added advantage is that the virtual model can easily be modified, adjusted, transformed, machined. On the other hand, the drawing remains most important after the design phase. It is a convenient way of conveying the information. It remains an excellent vehicle for the information pertaining to the manufacturing phase.

CADMAC System

The work described concerns the production of software system which would allow mechanical component designers to produce drawings in much shorter times than by existing manual methods and lower cost [1 6]. The software allows the user to create a library of previous and new designers which form part of a larger data base.

Component designs can be quickly built up in the computer using MENU commands to input and position lines, circles and other symbols. An interesting feature is the system library which permits high-speed design information retrieval from a data base. Standard components such as spacers can be retrieved from the data base and tried in an assembly. The component can then be included in the design or rejected, or another component tried at the discretion of the designer. It is possible to try all the relevant existing designs in the data base, and to prove that a new design is necessary should this be the case.

The COMPAC 3D CAD system uses an input language similar to APT [15]. Once component dimension data has been sotred, the system can produce graphical outputs showing workpiece surface contact, surface penetration, sections, and removal of hidden lines, see Fig.11. Also, modifications to workpieces and movement simulations can be done interactively. The system can be expanded to cope with the treatment of complex surfaces not amenable to mathematical analysis.

Die mould and pattern design ,can be considerably reduced by using computer see Fig.12.

CONCLUSION

This work present a comprehensive views for all the important aspects of CAD techniques and its application to mechanical design. It is believe that the use of interactive CAD system may increase productivity in engineering design as much as tenfold. It is also shown that CAD represent an excellent tool by which the engineering design process,data,analysis, optimization eliminates a great many of the manual steps.

The computer has made it possible for engineering to very accurately produce drawings. The introduction of computer graophics to the productio design environment has exposed many hardware,software,administrative and organizational problems which are limiting factors to productivity,and which are continually subjec to partial corrective action. Advances in computer technology have opened the doors to CAD , CAM,computer data base,etc

REFERENCES

1. Groover,M.P and Zimmers,E.W"CAD/CAM:computer Aided Design and Manufact-uring", Prentice Hall , 1984.

2. Smith,W.A, "A Guide to CADCAM", Pub. by I.Prod.Eng.& NES,UK, 1983.

3. Spur,G, Krour,F.L, and Harder J.J." The compac solid model" Computer in Mech.Eng. (CIME),vol.1 Nc2 pp 44-53, USA (Oct. 1982).

4. Besant,C.B, "Computer Aided Design and Manufacture"Ellis Horwood,UK,1980

5. Krouse,J.K," CAD/CAM"-Bridging the Gap from Design to production",. Machine Design,pp 117-125,(June 12,1980).

6. Boyse,J.W, and Gilchrist,J.E"GM Solids , IEEE,Computer graphics,pp 27-40 (March 1982).

7. Smyth,S.J."CAD/CAM Data Handling from Conceptual Design through Product Support, Jr. of air craft,pp 753-760,(Oct.1980).

8. Spur,G. and Germer,H.J." 3 Dimensional Solid Modeling Capabilities of the COMPAC System and Some Applications", CAE 82,Workshop in Geometric Moeling, Milan, (Feb, 1982).

9. Shigley,J.E."Mech. Eng. Design" 3 rd edition,McGraw-Hill Co.,N.W.,1977.

10. Leroux,D."Introduction to the Computer Aided Design System" R-A-3D" CAD in Medium Sized and Small Industries, MICAD., 1981.

11. Grabowski,H. " Preparation of Application of CAD - Systems for Economical Use, Ibd.

12. Schofield,N.A., and Smith,J.P.B " Soft Ware Considerations in the use of Computers for Design and Manufacturing Engineering, Ibd.

13. Rimbeanx,J. and Schneiser, A." CAD Pilot Project in Mechanical Engineering of Contraves AG in Zurich, Ibd.

14. MATRA DATAVISION, EUCLID , True CAD information, 8000 Munchen 80 West Germany (1983).

15. CAD and CAm Come Together , Jr. of Production Eng. pp 49-52,UK(June1981).

16. Jebb,A.,C.B.Besant and R.C.Edney,"The Application of CAD Techniques to Machine Tool Component Design, Production and Manufacture, 15th MTDR conf. procd. pp115-119,Birmingham,UK (1974).

17. National Engineering Laboratory (NEL),The Computer as a link between Design and Manufacturing , East Kilbride ,Glasgow, G75 OQU, UK.

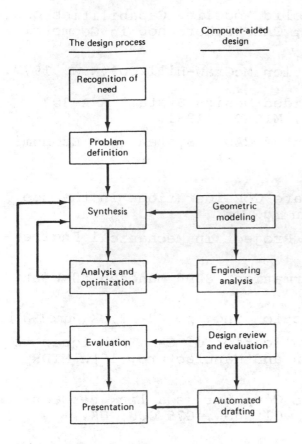

Fig.1. Application of computers to the design process.

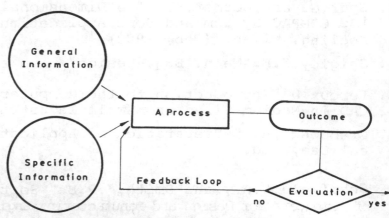

Fig.3. Diagramatic presentation of design iteration.

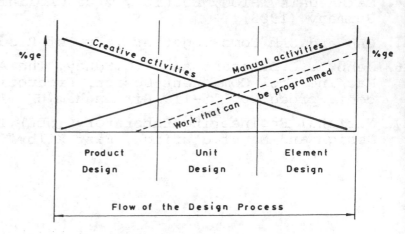

Fig.4. Activities in design phases.

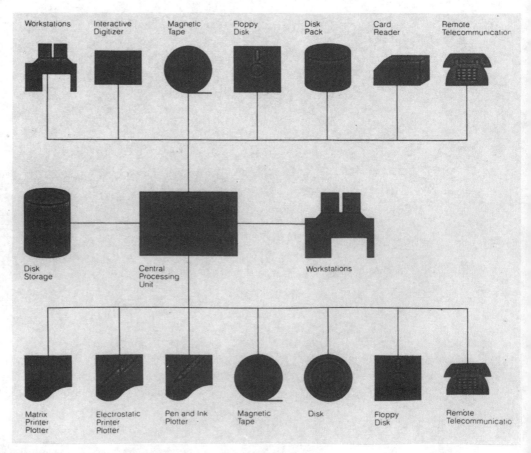

Input
Graphics data (manually created drawings and sketches) is converted into computer data via a digitizer. After raw data has been converted to computer data, input can be from a number of sources.

Process
Data is processed in the Central Processing Unit. Disk storage is used to store drawings and software.

Each workstation has its own processing capability permitting multiple workstations to share the same Central Processing Unit.

Output
Graphics output can be produced from a number of hard copy devices and/or sent to a storage media.

Fig.5. Basic hardware structure of a digital computer.

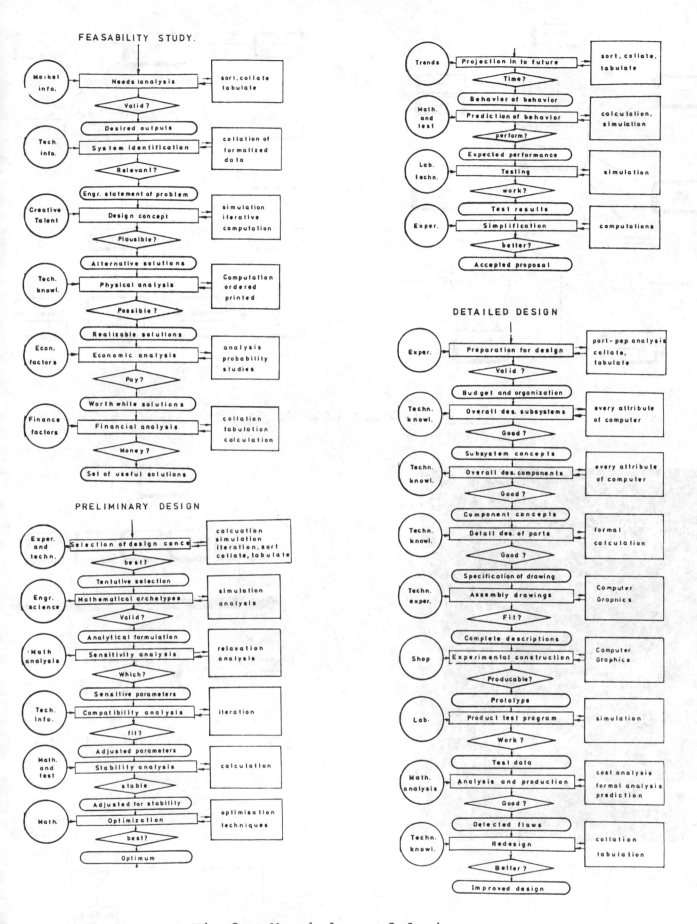

Fig.2.   Morphology of design.

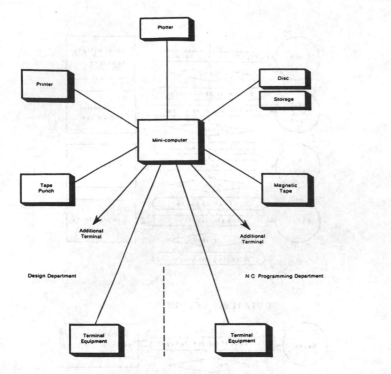

Fig.6-a

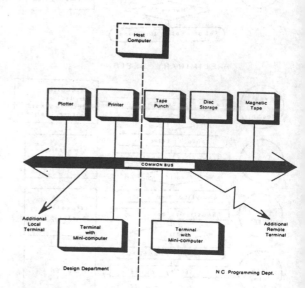

Fig.6-b

Fig.6. CAD and CAD/CAm System.

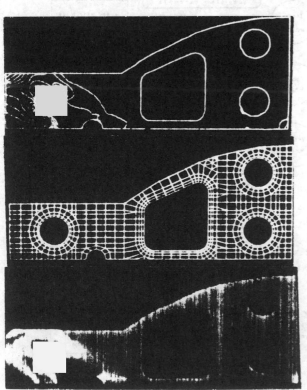

Fig.9. Typical finite element models.

Fig.7. CAD/CAM system networkin

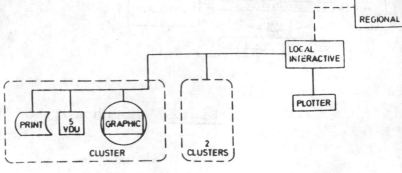

Fig.8. Developments in mini and micro computer.

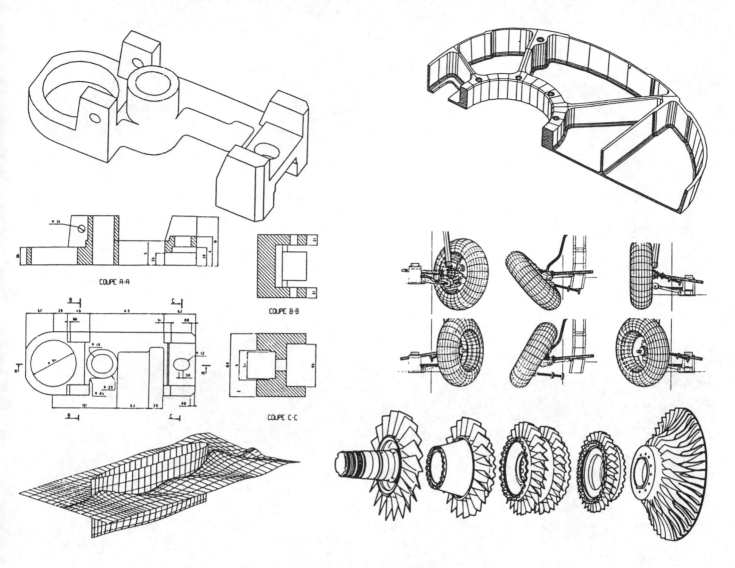

COUPE A-A

COUPE B-B

COUPE C-C

Fig.10. Different CAD examples.

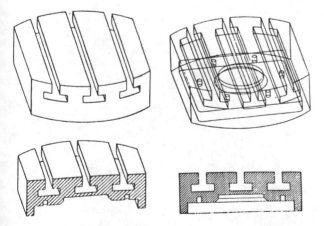

Fig.11. Graphic output and hidden line
remove in facility of the COMPAC,CAD system,

Fig.12. Model for die mould.

AUTOMATED PROCESS PLANNING IN FLEXIBLE MANUFACTURING SYSTEMS

Andrew Kusiak
Department of Industrial Engineering
Technical University of Nova Scotia
P.O. Box 1000
Halifax, Nova Scotia B3J 2X4 Canada

ABSTRACT

A considerable effort has been spent on developing automated tools for process planning. However, as yet there is no efficient automated process planning system available. This is a result of lack of appropriate methodologies for solving such a complex problem.

In this paper a new modeling approach to this problem and some of the solution procedures are presented. A role of the automated process planning in flexible manufacturing systems is also discussed.

INTRODUCTION

The traditional link between Computer Aided Design (CAD) and Computer Aided Manufacturing (CAM) is process planning. Conventional process planning has been done manually. It has been a tedious and ineffective activity based on empirical data, previous experinece, tables of standard times, tool catalogues, cutting diagrams, etc. (see Vernadat and Graefe 1983).

Currently there are many CAD and CAM systems in use around the world. It is natural that the link between these two systems as a highly Automated Process Planning (APP) system should be developed.

Bjorke (1967) was one of the first researchers who created the theoretical basis for an APP system. He applied topology and graph theory to the part description. To date a large number of researchers and practitioners have been working on the problems concerned with the APP systems. Takano et al. (1975) surveyed some of the APP systems. Beeby and Thompson (1979) presented a concept of an APP system for Boeing Commercial Airplane Corporation. Chang et al. (1982) listed some of the commercially available APP systems such as: APPAS (USA), CAPP (USA), EXAPT (West Germany) and

61

GE (USA). They also presented a new concept of an APP system. Emerson and Ham (1982) described an APP system named ACAPS, designed for rotational parts. A system known as SAPOR-S developed for NC-lathes is discussed by Gatolo et al. (1983). Yellowley and Kusiak (1984) presented a mathematical programming approach to process planning. Kusiak (1984) discussed a number of modeling approaches to process planning in the flexible manufacturing environment.

There are two basic approaches to the design of APP systems (Houtzeel 1981).

a) variant process planning based on the retrieval of standardized (e.g. group technology) process plans previously developed. The efficiency of this approach depends first of all on the quality of standardized information stored.

b) generative process planning generates the process plan from scratch based on the information stored in a CAD data base. In a pure generative system a process plan should be generated without any human intervention.

Most of the currently available APP systems are of variant nature. Some of them may also use generative logic, for example the system presented by Emerson and Ham (1982).

In this paper a hierarchical appraoch to the APP system will be presented. A fully generative APP system can only be created on the basis of a mathematically formalized planning process. Because of the nature and the complexity of this problem optimization theory, as inidated in Kusiak (1983), and pattern recognition will play a dominant role.

In this paper four process planning models are discussed. Each of them may arise in practice. The first three models are based on the set partitioning problems, and the fourth one on the traveling salesman problem. The advantages and disadvantages of both approaches are discussed. The final part of the paper deals with the solution methods of the four models. Detailed integer programming formulations of the four models are presented in Kusiak (1984).

FMS REQUIREMENT ON A PART PROCESS PLAN

As described in Kusiak and Wilson (1983) scheduling flexibility is one of the most important components of the overall FMS flexibility. The scheduling flexibility can be measured by the number of different routes along which a given job can be manufactured (Buzacott 1983). Process planning has a direct impact on the scheduling flexibility, because for a given part routes correspond to the variants of a process plan. In the classical manufacturing system, typically, an assumption was made that for one part only one process plan was available. This assumption is not valid in the FMSs, where it is required that one part has alternative process plans available. This is due to scheduling flexibility.

The scheduling flexibility in an FMS imposes the following requirement on the APP system:

1) it should ensure a dynamic generation of process plans with alternative machining routes. This is due to the elimination of idle time of machines while scheduling parts.

2) there should be a way to dynamically generate process plans with prespecified numbers and types of tools. Machine tools in FMS may have limited tool magazine capacity. Moreover, some of the tools may break down during the machining process. This in turn creates additional demand on space in tool magazines. To avoid changeover of the tool magazine sometimes it is required to restrict the number and type of tools used for maching of parts.

In the practice of process planning there are many routes in which a particular part can be manufactured. The number of routes is determined by:

- part complexity
- availability of tools
- shape of the initial material.

This paper is concerned with the optimization of process planning. The optimal process plan determines the part orientation. In some cases the optimal phase may not be accepted, because it would result in technologically infeasible part orientation.

MODEL M1

This model is developed under the following assumptions:

1) there is no limit to the number of tools which can be used to remove material from a manufactured part

2) there is no cost for tool usage

3) costs of removing material volumes are sequence independent

4) there are no technological constraints concerned with volumes to be removed.

In order to formulate this model consider a part in Figure 1 with a set $\{V_1, V_2, V_3, V_4\}$ of material volumes to be removed.

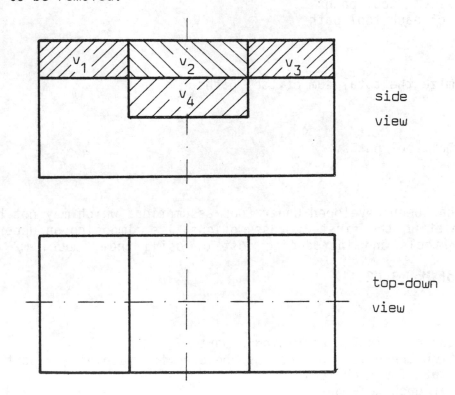

Figure 1. Part with material volumes V_1, V_2, V_3 and V_4 to be removed.

For a part in Figure 1 one can construct an incidence matrix (1). Each row i in the following incidence matrix A represents the volume to be removed. Each column V_j of (1) represents an admissible tool path at cost c_j. For example in the 5th tool path volumes V_1, V_2 and V_3 are removed at cost c_5. Each column of the matrix corresponds to a feasible tool path.

In a process plan there is a set of tool paths such that each volume is removed exactly once. An optimal process paln is a process plan with minimum corresponding costs.

Set of admissible tool paths

$$A = [a_{ij}] = \begin{array}{c} \\ \end{array} \begin{array}{ccccccc} 1 & 2 & 3 & 4 & 5 & 6 & 7 \\ \left[\begin{array}{ccccccc} 1 & & & & 1 & & 1 \\ & 1 & & & 1 & 1 & \\ & & 1 & & 1 & & 1 \\ & & & 1 & & 1 & \end{array}\right] & & & & & & \\ c_1 & c_2 & c_3 & c_4 & c_5 & c_6 & c_7 \end{array} \quad \begin{array}{l} V_1 \\ V_2 \\ V_3 \\ V_4 \end{array} \quad \begin{array}{l} \text{Set of} \\ \text{material} \\ \text{volumes to be} \\ \text{removed} \end{array} \qquad (1)$$

Based on the above considerations model M1 will be formulated. From matrix (1), the associated optimization problem can be formulated as the set partitioning problem.

Formulation of Model M1

Input

- volumes to be removed
- admissible tool path
- cost of each tool path

Objective

- minimize the total sum of tool paths

Output

- optimal tool paths.

MODEL M2

Model M1 has been developed under four assumptions which may not hold in many applications. Deleting the first two assumptions, i.e. imposing an upper limit on a number of available tools and introducing cost of using them, model M2 can be formulated.

Formulation of Model M2

Input

- available tools for machining a part
- limit on the number of tools to be used for machining a part
- tool path for each tool
- cost of each tool path
- utilization cost of each tool

Objective function

- minimize the total sum of cost of tool paths and utilization of tools

Output

- optimal tool paths and optimal number of tools.

MODEL M3

Both models M1 and M2 generate only an optimal set of tool paths. A new problem of sequencing these tool paths arises. This problem can be formulated and solved based on general sequencing theory. At least two excellent references, one by French (1982) and the other by Baker (1974), entirely on sequencing and scheduling can be indicated.

The sequencing of tool path problems can be formulated as a single machine problem with technological constraints. These constraints can be very easily represented graphically. To show the graphical reporesentation of the technological constraints consider a part shown in Figure 2.

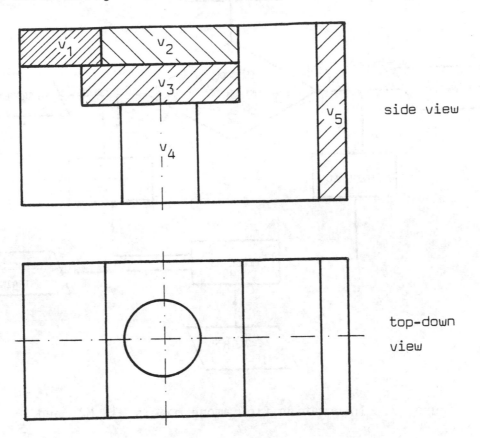

side view

top-down view

Figure 2. A prismatic parts with volumes V_1, V_2, V_3, V_4, V_5 to be removed

Assume the following tool paths generated:

$$r_1 = \{V_1, V_2\}$$

$$r_2 = \{V_3\}$$

$$r_3 = \{V_4\}$$

$$r_4 = \{V_5\}$$

The technological constraints can be expressed by the following digraph

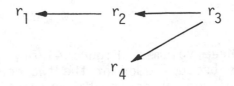

where $r_i \longleftarrow r_j$ reads, tool path r_i preceeds tool path r_j.

65

The two models M1 and M2 can be incorporated with the sequencing model M3 as presented in Figure 3.

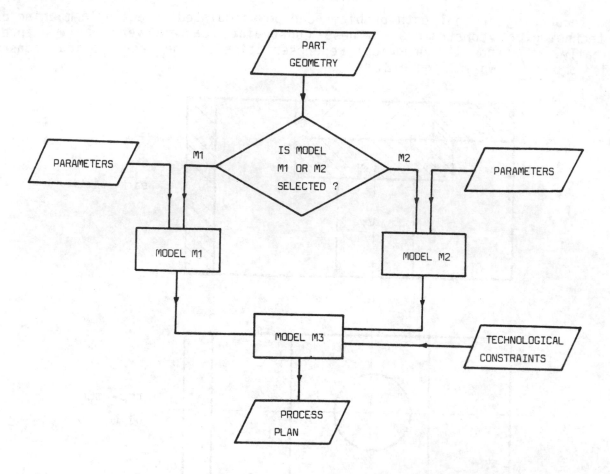

Figure 3. Information flow among models M1, M2, and M3.

The flow of information presented in Figure 3 is based on a part geometry data, model M1 or model M2 is selected. Solution to the selected model, together with technological constraints is input to model M3. Solving model M3 the optimal process plan is generated.

MODEL M4

In this model all the assumptions of paragraph 2 are relaxed. Such a case may occur in practice, if a cost of removing of volumes of material depends significantly on their sequence. Figure 4 presents a prismatic part with volumes V_1, V_2, and V_3 to be removed.

There are six sequences of removing the material volumes is Figure 4 shown below.

$\{V_1, V_2, V_3\}$ $\qquad\qquad\qquad$ $\{V_1, V_3, V_2\}$

$\{V_2, V_1, V_3\}$ $\qquad\qquad\qquad$ $\{V_2, V_3, V_1\}$

$\{V_3, V_1, V_2\}$ $\qquad\qquad\qquad$ $\{V_3, V_2, V_1\}$

The cost of removing these three volumes (Figure 4) may be different for each of the above six sequences. This is because each of the two double crossed material volumes in Figure 4 can be removed, either in horizontal or vertical tool path. This in turn may be associated with different removal costs.

The problem of removing material volumes with sequence dependent removing costs

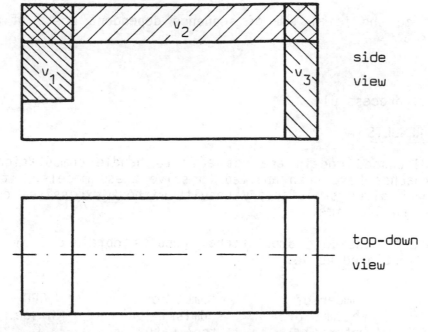

side view

top-down view

Figure 4. A prismatic part with material volumes
V_1, V_2, and V_3 to be removed

can be formulated as the traveling salesman problem (TSP). For the problem in Figure 4, a matrix S of a sequence of dependent costs can be formulated as follows:

$$
S = [s_{ij}] =
\begin{array}{c}
\begin{array}{ccc} V_1 & V_2 & V_3 \end{array} \\
\left[
\begin{array}{ccc}
\infty & C_{12} & 0 \\
C_{21} & \infty & C_{23} \\
0 & C_{32} & \infty
\end{array}
\right]
\begin{array}{c} V_1 \\ V_2 \\ V_3 \end{array}
\end{array}
\qquad (2)
$$

It is obvious that in matrix (2) the entries s_{13} and s_{31} are equal to zero because removing of volumes V_1 and V_3 does not depend on the removing sequence.

There is one more aspect which should be considered in the formulation of model M4, namely the initial and final position of the tool. Before the part machining starts a tool has its initial position. A cost of moving the tool from this position to the first material volume be removed usually depends on a volume number. Also, the cost of moving the tool from the last material volume removed to the tool final position depends on the number of the last volume removed.

A detailed analysis of the tool initial and final state is presented in Kusiak (1984).

Formulation of Model M4

Input

- volumes to be removed
- sequence dependent costs of removing any two subsequent volumes
- preselected tools for removing all the volumes

Objective

- minimize the total sum of sequence dependent costs of removing volumes of material

Output

- optimal process plan.

COMPUTATIONAL RESULTS

All four discussed models are not easy to handle computationally. A number of different approaches have been applied to solve these models. It appears that a sub-gradient approach gives satisfactory results without excessive requirements on computing resources and CPU time.

As an example consider computational results obtained from solving model M2 by the algorithms mentioned above.

Number of volumes of material	Number of admissible tool path	CPU in sec.
10	100	2.5
20	200	8.2
20	200	9.3

All the computation has been done on a CDC CYBER 170-720 computer. The presented results are 10% from optimality. More detailed discussion on the algorithms and computational results can be obtained from the author.

CONCLUSIONS

This paper discusses a very important aspect of process planning. The models presented hopefully will contribute to the development of a fully APP system. Availability of an APP system becomes extremely important in FMSs because of scheduling flexibility which can be measured by a number of alternative process plans available for each part.

ACKNOWLEDGEMENTS

This research has been partially supported by the Natural Sciences and Engineering Research Council of Canada.

Many thanks to Gail Veinot for very rapidly producing the typescript to her usual exacting standards.

REFERENCES

1. Baker, K. R., 1974, Introduction to Sequencing and Scheduling, John Wiley, New York.
2. Bjorke, O. and Haugrud, B., 1967, Mathematical methods in planning of machining operations, Presented at the CIRP Meeting, Ann Arbour, Michigan.
3. Beeby, W. D., and Thompson, A. R., 1979, A broader view of group technology, Computers and Industrial Engineering, Vol. 3, pp. 289-312.
4. Buzacott, J. A., 1983, The fundamental principles of flexibility in manufacturing systems. Proceedings of the 1st Int. Conf. on Flexible Manufacturing Systems, Brighton, U.K., pp. 13-22.
5. Chang, T. C., Wysk, R. A. and Davis, R. P., 1982, Interfacing CAD and CAM - A study in hole design, Computers and Industrial Engineering, Vol. 6, pp. 91-102.

6. Emerson, C. and Ham, I., 1982, An automated coding and process planning system using a DEC PDP-10, Computers and Industrial Engineering, Vol. 6, pp. 159-168.
7. French, S., 1982, Sequencing and Scheduling, John Wiley, New York.
8. Gatalo, R., Rekecki, J., Borojev, L.J., Zeljkovic, M., Milosevic, V., Konjovic, Z. and Malabaski, D., 1983, Automatic design of the technological process for NC lathes by the use of SAPOR-S system, International Journal of Production Research, Vol. 21, pp. 197-213.
9. Houtzeel, A., 1981, Computer - assisted process planning minimizes design and manufacturing costs, Industrial Engineering, Vol. 13, pp. 60-64.
10. Kusiak, A., 1983, An automated system of mechanical part recognition, Working Paper No. 12/83, Department of Industrial Engineering, Technical University of Nova Scotia, Halifax, Nova Scotia.
11. Kusiak, A., 1984, Process planning and scheduling problems in flexible manufacturing systems, Presented at the TIMS/ORSA Joint National Meeting, May 14-16, San Francisco, CA.
12. Kusiak, A. and Wilson, G.P., 1984, Recent developments in flexible automation, Proceedings of the 98th Congress of the Engineering Institute of Canada, May 20-25, Halifax, pp. 45-55.
13. Vernadat, F. and Graefe, P.W.U., 1983, Interfacing CAD and CAM, Proceedings of the 2nd Canadian CAD/CAM and Robotics Conference, Toronto, pp. 12.9-12.16.
14. Yellowley, I. and Kusiak, A., 1984, Observations on the use of computers in the process planning of machined components, Proceedings of the Annual Conference of the Canadian Society for Mechanical Engineering, Halifax, May 23-25, pp. 155-186.

ANALYSIS OF SHOP FLOOR TIMES IN BATCH PRODUCTION

Prof. Augusto De Filippi, Turin Polytechnic, Italy

Small and medium-sized firms have so far been able to make no more than
a marginal use of CAD/CAM, particularly on account of the shortage of
software suitable for microcomputers.
As a first approach to computerised production management, this paper
investigates two quantities of major importance for a manufacturing business, namely machining time and throughput (or flow) time.
It was found experimentally that the method used to predict the first
of these variable is very imprecise and optimistic, while throughput
times show over-high average values.
Prediction of flow time demands the use of a probabilistic mathematical
model. The frequency distribution of the ratio between this time and
the machining time was examined with a view to formulating such a model.
The results showed that a Weibull or an exponential distribution enabled
the experimental results to be interpreted with an excellent degree of
approximation.

1. INTRODUCTION

CAD/CAM techniques can be applied to advantage even by small and
medium-sized firms producing single or repeated batches of products,
since currently employed manual procedures for planning and checking
production, though established by long experience, nonetheless frequently lead to the following drawbacks:

- unpredictably long and chancy delivery times;
- continuous changing of programmes;
- numerous jobs completed behind schedule;
- a high volume of work in progress.

Many of the CAD/CAM systems available, however, in addition to being designed for large computers, are more concerned with questions related to design, or to tool path control of NC machines.

The advent of high-capacity micros, on the other hand, can now bring CAD/CAM techniques within the reach of small businesses. The problem of providing suitable software remains, of course, though Walker (cited by Steudel [1]) has shown that a generative planning programme employing an Apple II computer can be used to create process plans and NC tapes for a group of chucking machines.

As a first approach to the CAM aspect of this question, an experimental investigation of throughput time (or flow time) for batch production and its involvement with the manufacturing process is described in this paper. It was felt that evaluation of this quantity was an essential step in understanding the production process as a means of determining whether a job could be finished by the due delivery date, as well as forecasting the earlier delivery date.

Upline from this evaluation, however, there is a second extremely important time parameter, namely the machining time. This can be calculated in a variety of ways, ranging from the determination of the time allotted for each machining operation, to the use of statistical methods based on historical data [2].

The reliability of such techniques has also been checked experimentally by comparing the estimated times with the observed times.

2. EXPERIMENTAL PART

The experiments were carried out in two Turin firms (C and S). Firm C makes vertical and horizontal spindle surface grinding machines, and employs 35 clerks and 68 workmen. Firm S primarily manufactures gear shaving machines. Its workforce consists of 42 clerks and 80 workmen. Both firms use NC and universal machine tools. Their production batches (made for internal use) range in size from a few units to several dozen parts. Random samples of 50 parts for firm C and 52 for firm S were made up for the purposes of the experiment.

2.1. Machining time

This parameter is strictly composed of the preparation time, the actual cutting or machining time, and unproductive or idle time. To this must be added allowances for rest time occasioned by fatigue and for down time due to various contingencies. The total constitutes the estimated machining time.

This quantity is usually calculated by simple valuation in small and medium-sized firms. Tables exist for the rapid determination of preparation times and secondary times, from the breakdown of the machining operation as a whole into various elementary stages. This method is adopted by firms C and S. Its rapidity is an undeniable advantage. As we shall see, however, it gives rise to predictions that are very approximate and very optimistic.

For the purpose of the experiment, the estimated machining time (t_p) was worked out from the forms prepared by each company's methods-time office.

The real machining time (t_r) was calculated from the dockets accompanying the pieces during handling through the workshop.

2.2. Throughput time

The throughput time (t_f) corresponds to the interval between the date of arrival of a batch J_i in the workshop and that of the completion of the final operation in its manufacture.

It is thus composed of the sum of the machining times and the waiting times within the workshop. This gives the following equation:

$$t_f = \sum_{j=1}^{g_i} p_{ij} + \sum_{j=1}^{g_i} w_{ij} \tag{1}$$

where:

g_i number of operations required for J_i;

p_{ij} machining time for the jth operation;

w_{ij} waiting time for the jth operation.

Evaluation of t_f is difficult a priori owing to the dependence of the waiting times on a variety of factors, e.g. the number of batches in the workshop, priorities, machine loads, etc.

A deterministic mathematical model has been proposed[3] for t_f, in which:

$$t_f = a K^b \tag{2}$$

being:

$$K = n^{0.5} \left(\sum p_i + (p_{max} - p_{min}) \right) \tag{3}$$

a, b positive constants

where:

n number of operations required to machine the batch;

p_i machining time for the ith operation;

p_{max} machining time for the longest operation;

p_{min} machining time for the shortest operation.

The t_f values for each batch were taken from the machining dockets to check the reliability of model (2), and determine which variables have the greatest influence on this parameter.

3. RESULTS AND DISCUSSION

A linear pattern is obtained with a good degree of approximation when t_r is plotted against t_p. The interpolation lines have the following equations, in which R is the correlation coefficient and times are expressed in min/pc:

$$\text{Firm C} \qquad t_r = 6.57 + 1.30\, t_p \qquad R = 0.91 \qquad\qquad (4)$$

$$\text{Firm S} \qquad t_r = 7.01 + 1.39\, t_p \qquad R = 0.94 \qquad\qquad (5)$$

The interpolation line for firm C is shown as an example with the 95% confidence bands in fig. 1 (the experimental values have been omitted for the sake of clarity).

Statistical analysis of results reveals that neither the differences between the constants nor those between the slopes are significant. This means that these two totally independent firms employ substantially the same estimation procedures. It is also clear that the method employed is very approximate and optimistic, since t_r is from as much as 1/3 to 1/2 more than t_p. The repercussions on cost forecasting can readily be imagined.

As far as t_f is concerned, our aim was to see whether the mathematical model (2) was applicable to these firms. It can be seen in fig. 2 (relating as an example to firm C) that the values are very scattered, which means that the deterministic model will not give reliable information.

In view of the gross uncertainty surrounding waiting times, it is clear that only a probabilistic model can provide proper simulation of the system. It was therefore decided to analyse the frequency distribution of the ratio between t_f and t_r, since the second of these quantities is the only one available for the probabilistic evaluation of the second In our opinion, no substantial refinement of the model can be obtained by looking for further correlations (e.g. including the number of operations), owing to the great inequality between t_r and waiting times. It will also be remembered that according to Phillips[4] the reciprocal of the mentioned ratio represents the efficiency of the production system.

The frequency distributions of the ratio between t_f (days) and t_r (hours) are plotted for firm C and firm S in figs. 3 and 4 respectively. The cumulative frequency distributions are shown in fig. 5. Several remarks may be made with respect to these figures. In the first place, the t_f/t_r ratio is much the same in each case (see fig. 5), even though firm S has a wider dispersion of values. It will be recalled that both firms had much the same t_r pattern. The most interesting feature, however, is the magnitude of the ratio: as a matter of fact the mean value is about 3.7 in both cases (leaving out one very high t_f for firm C, which was probably due to an abnormal manufacturing situation). As to the type of distribution, firm C may be supposed to have a Weibull distribution. Its density (if $x > 0$) is given by:

$$f(x) = \alpha \beta^{-\alpha}\, x^{\alpha-1}\, e^{-(x/\beta)^{\alpha}} \qquad\qquad (6)$$

The shape and scale parameters (α, β : both > 0) were determined from the experimental results:

$$\alpha = 1.54 \qquad\qquad \beta = 4.02 \qquad\qquad (7)$$

The probability values calculated with (6,7) are indicated by the broken line in fig. 3. There is a close agreement between the experimental and the calculated distribution. The chi square test shows that the fit is good at the 1% level of significance.

The experimental distribution for firm S, on the other hand, is of the exponential type. Its density (if x ≥ 0) is:

$$f(x) = \frac{1}{\beta} e^{-x/\beta} \tag{8}$$

The scale parameter was calculated as:

$$\beta = 3.72 \tag{9}$$

The values given by expressions (8,9) are indicated by the dashed curve in fig. 4. There is a clear agreement with the experimental and the theoretical data. Once again, the chi square test shows a good fit at the 1% level of significance.

Lastly, it should be noted that the difference between the two types of distribution is rather limited, since the exponential is nothing other than a particular case of the Weibull distribution (α = 1).

4. CONCLUSIONS

This study revealed two very serious drawbacks in the management of production at these two firms:

a) the method used to forecast machining times proved inaccurate in default, with the result that real times were under-estimated, together with their associated costs. The fact that this was a long-standing situation and common to both firms was indicative of a barrier between the methods-time office and the shop floor. Since there was no feedback of the out-turn data, this barrier prevented the method itself from being adjusted.

b) As far as t_f was concerned, waiting times occasioned by various forms of lack of organisation meant that the workpieces spent enormously long periods in the workshop. On an average, each hour of machining was accompanied by a flow time of nearly four days. This obviously gives rise to a very large amount of work in progress, leading to higher costs and delays in shipment.

Economically sound management of production must pursue three main objectives:
- short and exact delivery times;
- high utilisation of labour and machine tools;
- low production control costs.

Computers are an essential aid towards the attainment of these objectives. Suitable simulation programs can be used to provide dynamic solutions to machine load management. In addition, interactive scheduling methodology rather than complex job shop planning programs can be employed to handle the production process in small and medium-sized firms.

The experimental data were collected by Mr. F. Gay for his degree thesis. The writer, who acted as his supervisor, wishes to thank him for the important aid.

REFERENCES

1. Steudel,H.J., "Computer-aided process planning: past, present and future", Int.J.Prod.Res., 1984, nº2, p.253.

2. Cochran,E.B., "Using regression techniques in cost analysis. Part I and II", Int.J.Prod.Res., 1976, nº4, p.465, 489.

3. Barausse,G., "Programmazione automatica della produzione con micro-computer in una azienda meccanica", La Meccanica Italiana, May 1976, p. 15.

4. Phillips,K., "Aspect of job scheduling", Journal of Engineering for Industry, February 1979, p. 17.

5) Law,A.M., Kelton,D.W., "Simulation modeling and analysis", Mc Graw Hill, New York, 1982.

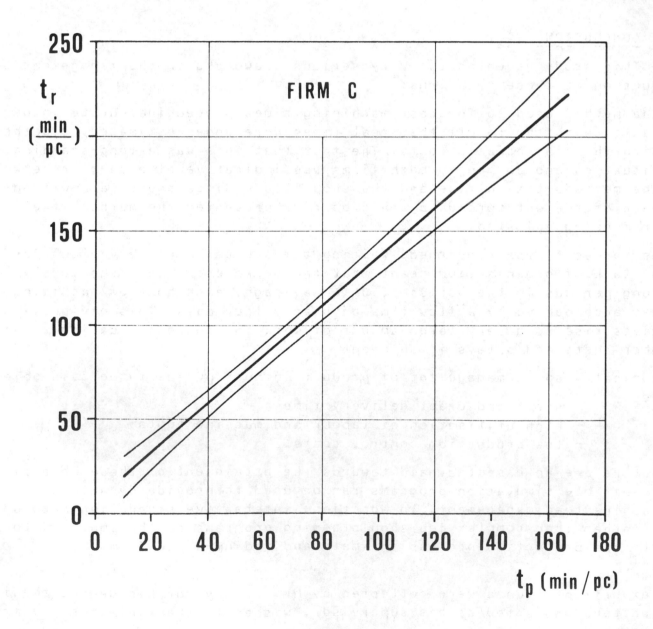

Fig. 1 -

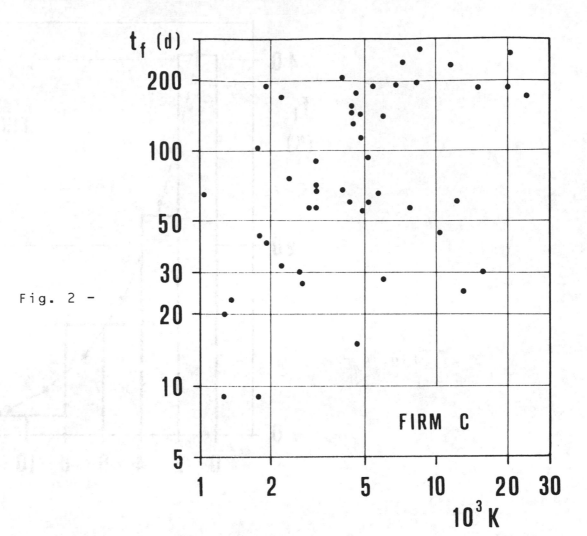

Fig. 2 -

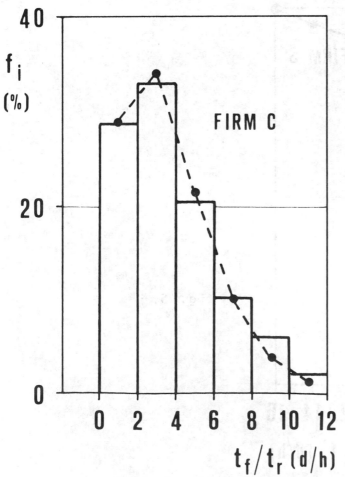

Fig. 3 -

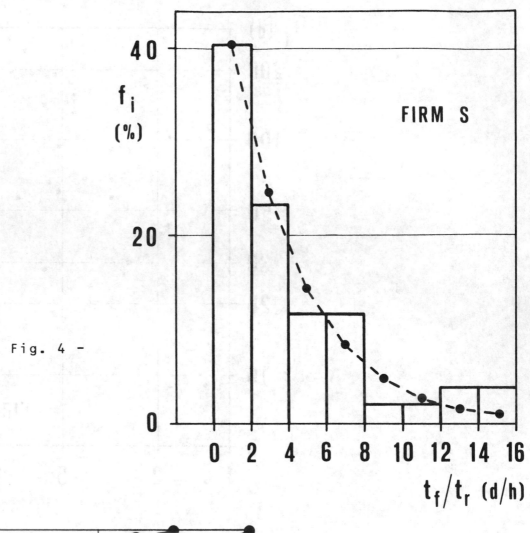

Fig. 4 -

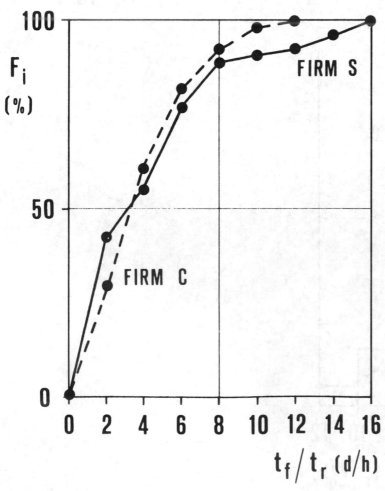

Fig. 5 -

COMPUTERISED SYSTEM FOR MANAGEMENT OF A TOOL ROOM : A CASE STUDY

S.R. SARDESAI, A. SUBASH BABU AND S. SOMASUNDARAM

DEPARTMENT OF MECHANICAL ENGINEERING,

INDIAN INSTITUTE OF TECHNOLOGY, BOMBAY, INDIA.

This paper pertains to the study carried out in a large 'Tool Room' of a multi-plant, multi-product manufacturing Organisation. The underlying objectives of the study were primarily to reduce the expected throughput time of the process of making of different tools. This study also included in its perspective, the other important problems of optimally allocating resources and working out efficient procedures for overall planning and control of the highly demanded tool room. This paper reports the details of the development of a computerised system using simulation methodology, whichwill also include the relevant details of the diagnosis and inferences of the case study.

1.0 INTRODUCTION

In manufacturing organisations, the production rate and quality of product depend very much on the quality of the tools; and hence the tool room plays a major role in achieving the targets of the production units. Any delay in tool making has direct bearing on the output of the production units. The process of tool making is a high precision and non repetitive type; and, in general, the facilities in a tool room include high precision machine tools and skilled labour which make the cost of operation high.

This paper brings out certain important issues involved in planning of a tool room while considering the case of the tool room of a multi-plant, multi-product manufacturing organisation. The paper includes discussion on the problems faced in the existing system and also the approach proposed to circumvent these problems.

2.0 LITERATURE REVIEW

Due to high precision machining operations, manual fitting and intricate and delicate assembly operations the process of tool making poses complex problems, especially with the point of view of planning. Also in spite of its importance in production, the tool room is often depicted as a service centre. Possibly these considerations have restricted the work of systems analysis in the area of tool room planning for a long time and hence the literature available is admittedly rather very much limited.

Steele [7] proposes a computerized system for estimating and manufacturing control, essentially with the idea of reducing the complexities caused by the uncertainties associated with the arrival of requests, work content, time taken and the material required. The steps in the proposed control procedures are

  i) design and introduction of computerized estimating
 ii) introduction of simple system of computerized labour-cost
     control
iii) computerized production and stock control.

A system suggested for computerized estimates for tool room [9] rests on acceptance that the work in tool room has high variability, which goes against any methodology requiring highly precision data of the type which is appropriate to any typical production system work. The statistical tool of multiple linear regression analysis is the foundation of the system developed. A standardized procedure is developed which is reported to be easy to apply to each new application, and thus to recalculate regression coefficients.

The above literature which is directly relevent to tool room planning, deals with estimating procedures. For the purpose of scheduling problem, it is realized that the problem of tool room is similar to the problem of multi-project scheduling, as each tool is made up of a number of components to be processed with certain technological ordering. A number of papers dealing with multi-project scheduling can be found in the literature. Fendley [3], Wiest [8], Pritsker [6], are few of those who have discussed various techniques in detail. It is felt beyond scope to present these techniques here. However, Davis [1], Graves [5], Gelder [4], Deshpande and Subash Babu [2], have given excellent review of this literature.

3.0 THE CASE

Here the outline of the case is presented alongwith diagnosis of the present system. Pertinent details of the analysis are also furnished with the help of the data collected.

3.1 Outline of The Case

The tool room under study consists of four departments namely, design, planning, manufacturing and tool trial.

The order for the new tool is received by the design department. The due date for completion of the tool is decided jointly by the customer (production unit) and planning and manufacturing departments. The planning department estimates materials and time requirements with respect to each of the eleven work centres of the manufacturing department. The estimation is not done for individual components of the tool, but for the whole tool. Each work centre (including manual fitting and assembly section) has a number of facilities and a total of 113 men are working in this department. At any time, on an average, about hundred and twenty tools are in the system at various stages of completion. The loading of various facilities and sequencing of different components of different tools are done by the supervisors on the shop floor. Once the tool is completed, it is tested in the tool trial department and if any defect is identified, the tool is immediately reworked.

3.2 Preliminary Analysis

In order to review the existing planning system and to identify the drawbacks, historical data was collected and the performance of the tool room was analysed.

For determining the load on the system the job arrival pattern was examined. The data collected over a period of twelve months revealed that there are three major types of tools viz., (i) press tools, (ii) Moulds and (iii) Jigs and Fixtures; and the approximate product mix for these tools was observed to be 56%, 32% and 12% respectively.

The observations also pointed out that about 50% of the tools were not completed within specified due date of completion. The utilization of the facilities was also found to be poor.

As the requirements of different facilities differed for the three types of tools; the break-up of work centre wise requirements (average) for three types of tools was determined from the past data. The result is presented in Table 1.

The data regarding the shop capacity is presented in Table 2.

3.3 Relevent Observations

The analysis carried out helped to identify the certain short-comings with respect to the following.

(i) Due date: For the new tool, its due date has to be decided before its design; but there is no formal method to determine the existing load on the system as it is impossible to judge total load of so many tools manually.

TABLE 1
TOOLWISE REQUIREMENTS OF WORK CENTRES FOR THREE TYPES OF TOOLS (IN %)

WORK CENTRE	TYPE OF TOOLS		
	PRESS TOOLS	MOULDS	JIGS & FIXTURES
MATL. CUTTING	6.5	4.1	8.3
TURNING	6.5	5.3	4.9
CYL. GRINDING	5.6	4.5	2.6
SHAPING	6.0	4.2	8.1
MILLING	11.2	9.4	16.8
SUR. GRINDING	20.2	14.5	16.8
JIG BORING	9.5	12.6	10.1
DIE SINKING	0.0	9.5	1.3
SPARK EROSION	1.0	3.3	0.4
WIRE EROSION	3.5	0.8	0.3
MANUAL	30.0	31.7	30.4

TABLE 2
INFORMATION REGARDING WORKCENTRES

WORK CENTRE	NO. OF FACILITIES	NO. OF MEN
MATL. CUTTING	4	8
TURNING	6	7
CYL. GRINDING	6	9
SHAPING	2	3
MILLING	6	10
SUR. GRINDING	8	12
JIG BORING	13	21
DIE SINKING	2	4
SPARK EROSION	2	3
WIRE EROSION	2	3
MANUAL	33	33

TABLE 3
CAPACITY UTILIZATION OF CRITICAL WORKCENTRES

WORK CENTRE	EXISTING UTILIZATION (%)	UTILIZATION WITH ADDITION CAPACITIES (%)	
		ONE FACILITY AT SHAPING	ONE FACILITY AT SUR. GRINDING
MANUAL	74.6	100	86.3
DIE SINKING	43.9	58.9	50.8
WIRE EROSION	50.8	68.2	58.8
SPARK EROSION	47.7	64.0	55.2
CYL. GRINDING	33.0	49.6	42.8
JIG BORING	33.0	49.6	42.8

82

(ii) <u>Priorities</u>: The priorities are decided by the supervisors by their own discretion and these priorities are changing as the individual judgement of the situation changes. Hence assigning priorities appears to be absolutely subjective and arbitrary.

(iii) <u>Time estimates</u>: The time estimates made in aggregate form are used only to estimate the cost of tool and not for the purpose of scheduling and planning.

(iv) <u>Number of tools</u>: There are, on an average, about hundred and twenty tools in the system at any given time and monthly turnover is about fifteen to twenty tools. Due to the present manual scheduling system, only few tools, which are to be completed in the immediate future are considered at any time.

(v) <u>Allocation of resources</u>: A lot of variation is present in the utilization of the various facilities, leading to arbitrary over-loading and underloading. Besides no systematic procedure appears to be existing for assigning jobs on facilities.

(vi) <u>Throughput time</u>: The average lead time for the three types of tools works out to be 7.5 months, 10.3 months and 9.4 months for press tools, moulds and jigs and fixtures respectively.

3.4 The Scope of The Study

The underlying objective of the study was primarily to reduce the expected throughput time of the process of tool making. The limitations of the present method of data handling and subjective judgements are believed to be causing the problems listed above. Hence it was decided to develop a suitable system and procedure essentially for planning various operations and scheduling so as to achieve reduction in throughput time. However, it was also felt necessary to limit the sophistication of the "to be proposed-system" in such a way that major changes in the infra-structure is avoided.

4.0 THE APPROACH

The literature offered very little help to identify a suitable procedure. Multi-project scheduling procedures have certain scope to be applied to such problems, but could be of no use in the present case as at any time there are about hundred and twenty tools in the system, which, of course may pose still more complications.

Hence a suitable approach has been evolved, the salient features of which are briefly outlined in the following sections.

In practice, efficient tool rooms require two important aspects to be incorporated. They are

(i) maximum utilization of the capacity and
(ii) minimum throughput time for tool making.

The first objective leads to the problem of capacity planning while the second leads to the problem of scheduling.

4.1 Capacity Planning

As the three types of tools need different facilities in different proportions as depicted in Table-1, an optimal mix of the three types of tools be processed to achieve the objective. As the tool room is a service centre, the orders are dependent on the customer; so to achieve the objective, the facilities should be planned such that the long term orders of the three types of tools utilize the capacity optimally.

The processing costs for different facilities being different, the value of facility utilization is taken as the product of the cost of processing per unit time and the total time the facility is used and hence, the total value of capacity utilization is considered as the sum of these values over all facilities for all products.

A Linear Programming model is conceived to achieve the objective of maximizing capacity utilization, constraints being the capacities of individual facilities. For the long term order pattern, the capacity requirement for each facility is calculated using this technique. It is also possible to calculate the optimal mix of tools for a given capacity. The Model and its brief description are given in Appendix I. The results are given in Table 3. The results of sensitivity analysis are also given in Table 3.

4.2 Shop Scheduling

Once the capacity of the tool room is planned to serve the long term orders, the aim of scheduling is to utilize this capacity to the maximum and to keep the total lead time of tool manufacturing to the minimum. From the capacity planning model the capacity available at each work centre for these three types of tools is known. The scheduling of different tools is done such that the capacity allocation is done optimally over the three types of tools. The following sections give the salient features of the sechduling procedure evolved.

4.2.1 Assigning Priority:

While resolving the clash of resources requirement for different tools, various important aspects are considered to determine priorities for the tools. They are as follows:

(i) Urgent jobs: At any given time there are some tools in progress which are termed as 'Urgent'. Such tools get the highest priority at any work centre.

(ii) Due dates: A tool with earlier due date should have priority over a tool with later due date if all the other things are the same.

(iii) Work remaining: A tool with more amount of remaining work content has higher priority with other things remaining constant.

(iv) Jobs in progress: A tool which is being processed should not be disturbed by a new tool unless the new tool is 'Urgent'.

The tools in the system have been classified into five types. They are (i) Urgent jobs in progress, (ii) Normal jobs in progress, (iii) Urgent new jobs, (iv) Normal new jobs, (v) Jobs in system but not in progress.

To determine priority amongst the tools from each category a priority index (PRIONO) is calculated for each tool which, takes into account due-date as well as 'work-remaining'. One category of tool has priority over other in the order (i), (iii), (ii), (v) and (iv). Fig. 1 depicts the hierarchy in which priority is fixed.

4.2.2 Precedence Relationship: There are variations in the tool making operations and operation times at work centre. However the various operations done on different tools follow more or less the same pattern. Also in spite of the fact that a tool may visit the same work centre more than once for processing; for planning over a longer time horizon, all the operations at a work centre have been lumped together for the purpose of scheduling. Also the sequence of operations for a tool can be conveniently represented by a precedence relationship diagram as shown in Fig.2. Such lumping of operational times at a work centre is justified because the total load on a work centre is of prime importance in the scheduling procedure followed.

4.2.3 Manual Operations: In the case of manual operations like fitting at different stage and assembly, a single person is allocated for a tool.

4.2.4 Simulation Model: The time for completion of tool making in the existing system is fairly long and hence the in-process work is also large. This leads to many tools at various stages to be considered for scheduling. This leads to complexity which in turn affects the performance of the system. Besides it is also almost impossible to resort to any mathematical modelling or some manual procedures. Hence a computer based simulation model is considered for evolving suitable scheduling procedures.

The operational characteristics of the tool room are simulated on the computer. The total load on the system at the starting point of the schedule is known from the real data. The historical pattern of the arrival of new tool is used to generate new tools and the load on various facilities in the subsequent periods. For each period the priorities of different tools is calculated and the allocation of time at different facilities is done after considering the precedence relationship of each tool. The logic of the methodology is outlined systematically in Appendix II, which is of course self explanatory.

5.0 RESULTS AND DISCUSSION

By using the capacity utilization model, the critical facilities in the existing set up are found. It is observed that the low cost centres such as shaping and surface grinding are bottlenecks which lead to poor utilization of costly work centres. To overcome this, extra capacity is planned for these operations. Sensitivity analysis with the model indicates that by one additional facility at shaping and surface grinding work centres, the capacity utilization improves by 50% and 30% respectively as presented in Table 3.

The model is based on the percentagewise break-up of the work centre at each facility for three types of tools as depicted in Table 1. The variation of these figures is studied over a period of twelve months and standard deviation is calculated. The sensitivity analysis helped to realize that for a change equal to one standard derivation, the overall utilization of the plant changes by maximum 4.8% in case of wire erosion work centre. Similar observations can be made by further analysis of the results.

Regarding the simulation model for scheduling, the logic of the model has been tested for its consistency and validity. However, a detailed investigation is being carried out to verify its appropriateness for the case data and performance with the point of view of the objectives considered. The exhaustive results obtainable with the model are perhaps beyond the scope of this paper, besides due to certain confidentiality requirements documentation is of course avoided, especially considering its implications prior to verification.

6.0 CONCLUSION

The study reported in the preceding sections of this paper is an effective step towards the computerized management of the tool room under consideration. It has the potential to offer a wide ranging advantages like proper assessment of resource requirement - in terms of quantity and priority; optimal allocation of resources accross different jobs; assigning the deserving priorities; identification of the jobs to be considered during a stipulated time; evaluation of performance measures deriving certain completion reports etc. Its compatibility to interface freely with computer and real life system through the simulation methodology is perhaps one of the interesting and useful aspects of this study which can of course be extended further to realize the benefits at interactive computerization.

Admittedly this study may have certain limitations especially considering the fact that it is one of the earlier studies in the area of tool room planning. However, further investigations will be carried out to refine this study to make it more general for real life planning of such tool room.

APPENDIX I

CAPACITY UTILIZATION MODEL

The model is formulated as a Linear Programming problem.
Let C_i (i = 1,11) be the cost of processing per unit time at i^{th} work centre.
Let h_i (i = 1, 11) be the total hours used at i^{th} work centre.
Let H_i (i = 1, 11) be the total working hours available at i work centre.
Let X_k (k = 1, 3) be the total hours used by k^{th} type of tool.

Let for each type of tool a fraction f_{ki} (k = 1, 3; i = 1, 11) represents the fraction of the total time needed for k^{th} tool at work centre i. (which correspond to percentage figures of Table 1)

$$\sum_{i=1}^{11} f_{ki} = 1, \quad k = 1, 3 \qquad \qquad ...(1)$$

Thus the objective function can be defined as Max $V = \sum_{i=1}^{11} C_i h_i$

The constraints of capacity may be put as $h_i \leq H_i$

where $h_i = \sum_{k=1}^{3} f_{ki} X_k$ $\qquad \qquad ...(2)$

i.e. in effect the LP problem is

Max $V = \sum_{k=1}^{3} \sum_{i=1}^{11} f_{ki} C_i X_k$ $\qquad \qquad ...(3)$

such that $\sum_{i=1}^{3} f_{ki} X_k \leq H_i , \quad i = 1, 11$

APPENDIX   II

SCHEMATIC DIAGRAM OF THE MODEL

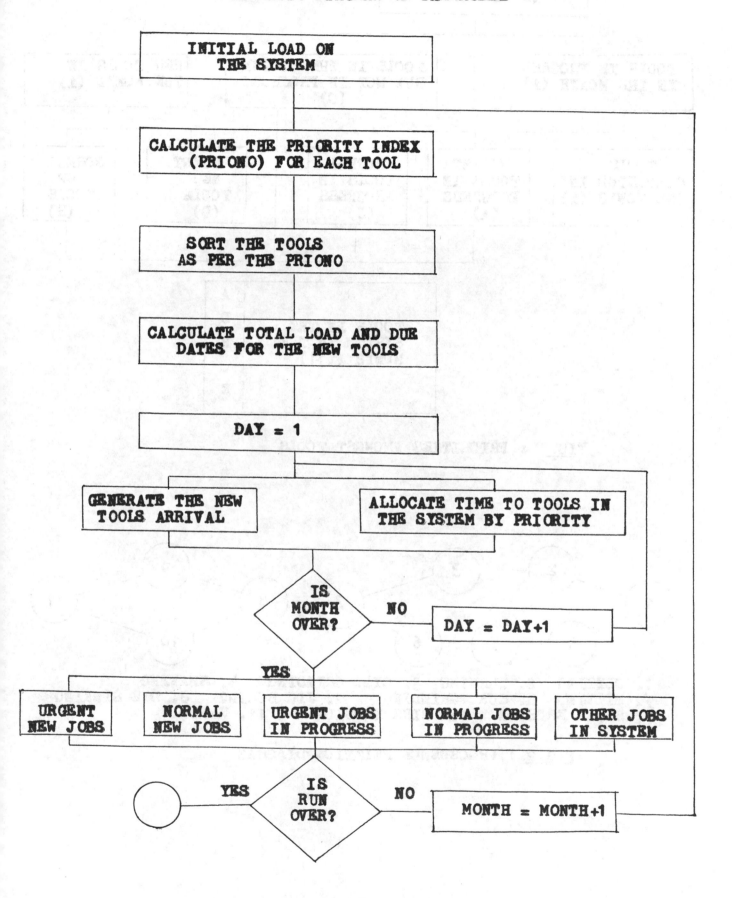

INITIAL LOAD ON
THE SYSTEM

CALCULATE THE PRIORITY INDEX
(PRIONO) FOR EACH TOOL

SORT THE TOOLS
AS PER THE PRIONO

CALCULATE TOTAL LOAD AND DUE
DATES FOR THE NEW TOOLS

DAY = 1

GENERATE THE NEW
TOOLS ARRIVAL

ALLOCATE TIME TO TOOLS IN
THE SYSTEM BY PRIORITY

IS
MONTH
OVER?

NO

DAY = DAY+1

YES

URGENT
NEW JOBS

NORMAL
NEW JOBS

URGENT JOBS
IN PROGRESS

NORMAL JOBS
IN PROGRESS

OTHER JOBS
IN SYSTEM

YES

IS
RUN
OVER?

NO

MONTH = MONTH+1

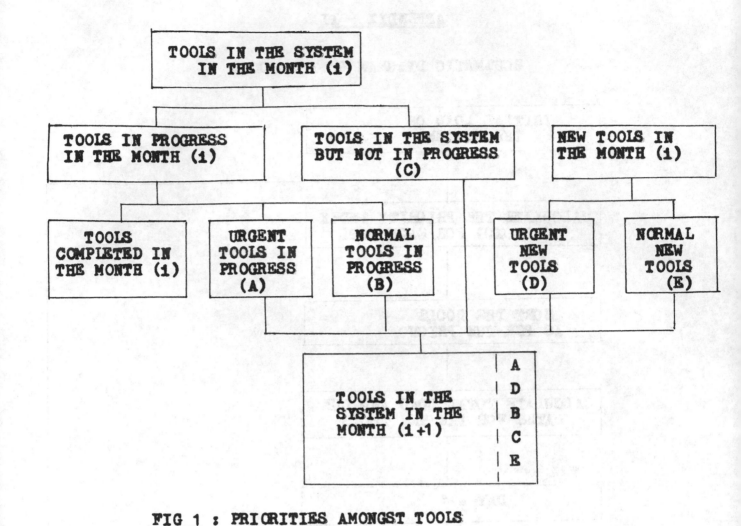

FIG 1 : PRIORITIES AMONGST TOOLS

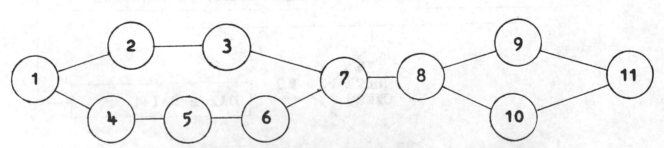

1. CUTTING 2. TURNING 3. CYL. GRINDING 4. SHAPING
5. MILLING 6. SUR. GRINDING 7. JIG BORING 8. DIE SINKING
9. SPARK EROSION 10. WIRE EROSION 11. MANUAL

FIG 2 : PRECEDENCE RELATION DIAGRAM

REFERENCES

1. Davis, E.W.   "Resources allocation in project network models: A Survey". Jnl. of IE, Vol. 17, No.4, pp. 177-180 (1966).

2. Deshpande, D.H.   and   A. Subash Babu.   "Multiproject multi-resources scheduling". Int. Cong. on Proj. Man. New Delhi (Nov.-Dec. 1983).

3. Fendley, L.G.   "Towards the development of complete multiproject scheduling system".   Jnl. of IE, Vol.19, No.10, pp.505-515 (1968).

4. Gelder, L.F.   and   Wassenhove, L.N.V.   "Production Planning : A review". Eur. Jnl. of OR, Vol.7, pp.101-110(1981).

5. Graves, S.C.   "A review of production scheduling". OR Vol. 29, No.4, pp. 646-675 (1981).

6. Pritsker, A.B.   "Multiproject scheduling with limited resources; A zero-one programming approach". Mgt. Sci. Vol.16, No.1, pp. 93-108 (1969).

7. Steele, P.M.   "Computerized control of tool room". Sheet metal ind., Vol. 59, No.5, pp. 465-472 (May 1982).

8. Wiest, J.D.   "A heuristic model for scheduling large projects with limited resources".   Mgt. Sci. Vol.13, No.6, pp. B 359-377 (1967).

9.   Anonymous   "No more guessing game in the tool room". Machinery and Prod. Engg., Vol.136, pp.23-25 (8 Feb. 1978).

REFERENCES

COMPUTER-AIDED MACHINING

INTERNATIONAL CONFERENCE ON
ADVANCES IN MANUFACTURING
9 - 11 OCTOBER 1984
SINGAPORE

Design elements, developmental tendencies and application of modern milling machines , machining centers and flexible manufacturing systems

Lutz Strakeljahn, Dipl. Engineer
MAHO Werkzeugmaschinenbau Babel & Co
Federal Republic of Germany

Within the entire range of manufacturing technology the trend today is towards higher main - drive performance, faster feed rates, higher rapid speeds and thus towards a reduction of non - productive time. This has been particularly influenced by the developments that are being made in the cutting - material sector at a sweeping pace and which go beyond hard metals and cutting ceramics to polychristalline diamond materials and thus place considerably higher performance demands on machine tools than was still the case ten years ago.

However it is furthermore of prime importance that due to the current existence of micro - processor technology on a wide - spread basis automation has been introduced to machine - tool control systems.

The Design of Modern Machine Tools

Frame (Ribbing, Guideways, Castings, Weldings)
Milling machines and machining centers for milling are fundamentally constructed of very rigidly ribbed basic elements, the column, the modular support elements, the headstock, the numerically controlled table.

These modular design elements are either made as castings, or according to the modern concept and in order to achieve high production - run ratings and with them a high degree of manufacturing

Due to the price decrease of micro - electronics more and more machines are now computerized numerically controlled and even today we can already forsee a time when virtually no more machines will be supplied without a numerical system.

In addition to control technology, monitoring technology is achieving increased importance. It enables machines to far surpass the previous productivity rate of 6% of the 8,200 hours available annually and to make use of 20, 40 even 60 or 70% of the available time.

This makes the machines decisively more productive, whereas on the other hand it causes a reduction of the number of machine tools that are absolutely required on the world market.

flexibility, they are made as welded constructios. The guideways are of decisive importance for the overall accuracy of the machines. Currently being used are cast guideways that are coated with nonmetallic compounds, this synthetic material is used because it prevents stick slips and also because advancements in die - casting technology have accounted for very advantageous means of producing it. Also being used even on production

91

machines are hardened steel guideways with synthetic linings. The guideways are superfinished milled, cast guideways with ceramic cutting material at speeds of 400 m per minute at feed rates of 2 m per minute, or hard steel guideways with CBN, or they are microfinished so that accuracy rates of 3 μm per meter are achieved.

Feed Drive

On modern CNC machines individual drives for all axes are used as feed drives. These are generally thyristor – controlled or transformer – controlled DC srevo drives, which are highly dynamic and are very easy to maintain. These motors provide the drive for each of the saddles by way of geared wheels, time belts and recirculating ball – screw systems.

Main Drives

Currently in place of the formerly predominating three – phase drives with 4 – step transmissions, DC drives are being used which provide for stepless speed regulation.

Achieving the high torques on the milling spindel which are needed for cutting operations usually requires 2 – to 4 – step reduction gears that are automatically switched, so that during speed changes the control can search for the optimal torque ranges.

Measuring Systems

On a precision machine the measuring systems must directly control the movements of the machine component that will be moved, if at all feasible without any intermediate links.

Therefore rotary resolvers that are mounted to the motor are less suitable for high – precision machining operatios than linear measuring systems which are mounted directly to the machine components and rotary measuring systems located directly on the table axis.

Control

Due to advanced integration of micro – electronics it is now possible to contain the complete controll unit in the control panel.

In addition to the central micro – processor, which processes all control data and the control keyboard, which is used to enter programs into the control but which can also be used to operate the machine manually, it is taken for granted these days that a modern control system is equipped with a visual display unit, which aside from information concerning the traverse position of the machine can also display any and all other relevant information to the operator.

The control concept displayed in the picture was developed with particular attention being placed on the aspect of "standard equipment control system" and thereby on the aspect of price merit.

This control is standardly equipped with a memory – capacity expansion of 54 Kbyte, a visual display unit and a diagnostic unit, which monitors the control as well as the most wideranging other machine components such as limit switches.

Furthermore the random – selection feature, which is required for machine centers is an integral component of the control and thus the control is suited for controlling machines with three axes, machines with four axes and machining centers as well.

It is not until the machining system is put into use for the first time that the control is informed by way of basic data that in this particular case it will be controlling a machine with 3 axes or on another case a machining center.

The simplification of servicing in which this results is enormous and as far as prices go hardly comprehensible.

Tables

For universal machines in addition to the simple rigid machining table it is also possible to use a universal rotary table with digital display. The turning and tilting movements of this table in all of the axes enable even the most complicated machining operations to be performed on the universal machine.

The highest degree of operational ease to be offered by tables is found with the numerically contolled rotary table, whose axes are completely numerically controlled and can be interpolated with the other axes of the machine so that spirals and other such shapes are made possible.

This table like the linear axes is driven by DC servo motors and by means of a high – reduction spur – gear transmission the feed motor speed is reduced to the lower table speed with the required high torque.

Total Concept

Basically modern CNC machines and machining centers should be constructed to form a compact unit and so that their overall dimensions enable them to be loaded onto a delivery truck and that once they have arrived at the customer's location, aside from certain preparatory measures concerning the setup of the machine, nothing more than an electrical hook – up will have to be performed to start the machine running.

The costs incurred during the installation of a new machine when the machine is delivered in various component parts and is not put together before arriving at the customer's location, where it is then put into operation again, very frquently go unrecognized.

Machining Centers

Machining centers perform basically the same milling jobs as milling machines. However these centers are primarily conceived for the production of average – sized to large batches. On machining centers the tools are automatically changed and the workpieces are automatically conveyed into the machining area.

Tool Changer

The tool changer displayed here is constructed in very strict accordance with analytical – value considerations and is in light of this high functional reliability extra – ordinarily reasonable in price.

The magazine plate for 24 tools is constructed of nibbled plate metal. The holders for the tools are made of spring steel and are equipped with high quality cam castings.

Whereas the movements of the changer arm are activated hydraulically, a Maltese – cross circuit is used to activate the magazin by means of a simple windshield – wiper motor which is very reasonably priced and when servicing is necessary is available everywhere.

Pallet Changer

In the case of the pallet changer as well, very strict attention was paid to functionability. The pallet changer consists of a relatively simple forklift, which brings the pallets into the machining area. All other movements necessary for pallet changing are performed by the machine axes. Since the pallet changer is equipped with just one drive, it too can be run very economically.

Pallet Storage System

The addition of a pallet storage system presents itself as a solution to making even better use of the production time of one or more machining centers, especially for work with heavy workpieces.

During regular production time the pallets of the storage system are loaded with workpieces at a suitable clamping station and the machining center transfers the pallets with the workpieces successively into the machining area and places the finished workpiece on its pallet back into the storage system.

Thus if all the pallets are loaded with a large number of workpieces at the end of a regular shift the machining center can then machine the workpieces in the subsequent time period without the intervention of an operator.

Enclosed Work Area

At the present time large amounts of coolants are used for work in machining centers. Thus 100 to 200 liters per minute are sprayed onto the workpiece through spray noozels, on one hand in order to cool and lubricate the workpiece and on the other hand to rinse away the large amounts of accumulated chips.

This makes a complete enclosure of the machining center a necessity, whereby this enclosure must be designed so that on one hand the operator retains visibility into the work area and on the other hand careful attention must be paid to the high quality of the sealing materials used (rubber, plastic etc.), as highly corrosive coolants frequently wear away at the sealing materials.

Universal Machines with Tool – Changing Equipment

As of recently, universal machines with their horizontal and vertical machining capabilities are also being equipped with supplementary tool – changing equipment, which to be sure works slower than the tool – changing equipment of machining centers but does not impinge upon the universality of the milling machine and, on the other hand offers a high productivity gain, since especially in low – staffed shifts the machines can then be used with the tool – changing system.

Developmental Tendencies

Increasing Main – Drive Capabilities and Increasing Demands on Accuracy

In addition to steadily increasing main – drive capabilities the demands placed on accuracy today are becoming constantly higher and higher and demands for higher productivity also continue to exist.

This higher degree of productivity has been made possible primarily by the fact that machines can now produce beyond the regular shift in second or even in third shifts, that is at times when the operating personnel is limited or is not present at all.

It is therefore an essential prior condition for low – staffed or unmanned manufacturing that certain monitoring functions, which are normally performed by the operator be carried out by automatic equipment.

Monitoring Equipment

For this reason a monitor of easily damaged tools seems to be of foremost importance. This type of tool – breakage monitor works by means of an optical system that checks the tool that has just been used and is now being changed back into the magazine for its rough dimensions, and whenever it ascertains that the tool is damaged, an alarm is sounded and the machine shuts itself down or a corresponding tool is selected.

Since fixtures represent a considerable capital investment, it is not efficient to use a pallet – storing system for machining exclusively one type of workpiece and thus to have the same fixture on all of the pallets. It is however much more economical when just one or two of the same type of fixture are used and the pallet – shunting system can then be loaded with various workpieces and a variety of fixtures. Accordingly each pallet must carry a code that clearly identifies the clamped workpiece and which summons the correct NC program on the control. Furthermore the coding of the pallet must be changed by the machine upon completion of the machining process so that the pallet is recognized by the machine when it makes another round and the workpiece on it is recognized as being already machined and is therefore not fed back into the work area again.

In order to achieve high rates of accuracy during the machining of workpieces, it is usually necessary to readjust the pallet - loaded workpiece inside the work area of the machine.

This readjustment can be done using a probe which is clamped into the milling spindel and then used to register certain reference points on the tool, whenever deviations from the reference data are registered, zero - point offsets in the three linear axes or in the rotary axis are performed by way of compensation.

In addition the probe can be used after machining has been completed in order to recheck the essential machining dimensions while the piece is still in work area of the machining center and to activate the appropriate alarm or to make program corrections if deviations from the reference data are registered.

The software - based monitoring of tool life is, to be sure not a completely exact means of registering the actual stage of wear and tear on a tool, but it is on the other hand very economical.

In addition to the regular tool data, such as length and radius, the expected life of the tool is defined in the control, and each time the tool is used, the cutting time is deducted from the programmed lifetime.

An alarm is actuated when the preset warning threshhold has been reached, and the tool can then be appraised or immediately replaced by the control personnel.

In order to take full advantage of the performance capability of the machines, while at the same time monitoring the wear and tear on the tools, cutting - power monitoring systems have been developed that, when intolerably high deviations from the reference cutting - power data due to increased wear and tear on the tools are registered, either activate an alarm or provide for the selection of a twin tool while at the same time influencing the feed so that the machine can be constantly travelled at its full performance rate.

Whereas machining centers are loaded with heavy parts by means of a pallet - storing system, light parts by comparison are being loaded into the work area of the machine at a steadily increasing rate by robot systems.

The workpieces are preoriented to the system as a whole and are placed in a corresponding workpiece storage system, the placing of the workpieces in the storage area is usually done manually since the

work involved here is minimal and the extraordinarily high flexibility of the human hand in conjunction with the human eye is required for this task.

The removal of the blank parts from the storage system is then performed by the robot, which also removes the finished parts from the work area of the machine and, if necessary brings them to a washing station or to a measuring station before finally placing them in the storage rack for finished parts.

Tool - Life Monitor

The easily operated contour control system assumes the task of tool - life monitoring. The permissible length of time that a particular tool can be used is assigned to it as its tool life.

Each time that particular tool is used the amount is added on to the total sum by the calculator of the contol system and when the treshhold has been reached, this is accordingly reported on the visual display unit and a corresponding spare tool, should one be available and called for is then used in all subsequent operations.

Tool - Breakage Monitor

For tools which are particularly susceptible to breakage, such as drill bits for example, the presence of the tool is monitored.

Should disturbances arise, tool breakage for example, the machining process is interrupted and a notice to this effect is displayed on the visual display unit.

When tool breakage occurs, the next tool is used, if available and called for. The tool - breakage monitor works according to the reflection principle where the transmitter and receiver are mounted on the same location.

The tool is checked in the tool magazine previous to being used and after it has been used. The bracket for measuring device is mounted to a stable axis of the magazine plate. The flap on the magazine hood is equipped with a reflector.

Cutting - Power Monitor (Motor - Current Monitor)

The cutting performance rates for each tool, which have been determined under production conditions or which are known through experience, are entered into the memory of the control.

Should this reference value be surpassed in the course of subsequent machining operations an overload alarm is sounded.

Probing System

A probe is removed from the tool magazine and is clamped in the spindle as a regular tool. It then travels to programmed measuring points on a given workpiece. The actual coordinates measured can be compared with the programmed reference values by the CNC control. Commensurate to the discrepancy value, zero - point offsets are caused in the three linear axes or in the rotary axis.

The probing system is particularly suited for the compensation of setup errors occuring during the clamping of the workpieces.

Multiple - Machine Operation

By the described methods of automatic workpiece loading together with such machine monitoring systems, multiple - machine operation is made possible, automatic operation in low - staffed or unmanned shifts and even in the so called graveyard shift has also been made feasible.

Thus operating personnel is steadily disappearing from production areas. The personnel thus freed is being used for the scheduling of manufacturing processes, for the creation of the part programs of the workpieces.

Workpiece programs are rarely created on the machines any more, the machines are still used to optimize the program when the workpiece is being machined for the first time, however instead workpiece programs are composed on external programming systems that are centered around economical personal computers consisting primarily of a keyboard, visual display unit, corresponding memory units and a plotter system for editing the part program.

Due to the performance ability of modern computers, more sophisticated internal computer languages can be used, which extraordinarily simplify and speed up the programming of the parts.

Application
Universal Machines

Today universal milling and boring machines are known throughout the world as high precision machine tools that can be universal applied. They are used in training centers, in tool and die manufacturing and in factory production ranging from small - batch to average - scale production.

They are used in the aeronautics industry, in some laboratories and even in nuclear - research centers, where due to radiation exposure, they are operated in hot cells by robots.

The basic machine is composed of machine body, column and headstock with horizontal and vertical spindels. Acessories which can be mounted to the face side of the headstock aside from the swing - away vertical milling head, which belongs to the standard equipment of the machine include an overarm, a high - speed vertical milling head, a slotting head, a multi - spindle head and an angular milling head.

The vertical clamping area of the column can be used as desired for mounting a rigid angular table, a universal built - in rotary table or an NC indexing

unit with face plate. With this modular system universal milling and boring machines are equipped for almost all jobs concerned with milling, boring, thread - cutting, slotting, grinding etc..

Tool and Die Manufacturing

It is precisely in the field of tool and die manufacturing that universal milling and boring machines are being used at a steadily increasing rate.

Here for example a very complicated workpiece can be machined from 5 sides in just one setup. This capability does not only reduce setup time, but guarantees a very high degree of manufacturing precision as well.

Typical machining examples in tool and die manufacturing are the manufacture of synthetic injection molds for such items as casing covers and the manufacture of cupped basins, closing sides etc..

Production of Prototypes

The contolled universal milling and boring machine is used for the production of single workpieces in prototype manufacturing in order to complete highly accurate metal - cutting jobs on several planes in just one setup.

Complicated contour runs are made possible on CNC universal milling and boring machines by the use of the 3 - axis simultaneous machining function, which is a standard feature of these machines.

Production of Average - Sized and Small Batches

Controlled universal milling and boring machines are being used at an increasing rate in the production of average - sized and small batches.

Program - controlled production runs and high repeatability precision and positioning tolerance of the universal milling and boring machine enable it to be used economically and efficiently in small - batch production.

By using multiple fixtures and accessories such as the multi - spindle vertical milling head and finally by equipping the universal milling and boring machine with automatic tool - changing equipment, it is possible to optimize the production of average - sized batches.

Machining Centers

The machining center is the result of continued further development based on the concept of the universal milling and boring machine.

In place of the vertical milling head, a tool changer was installed, which in conjunction with a tool magazine and by being CNC contolled provides for automatic performance of machining operations.

The chip - to - chip time during tool changing is less than 10 seconds. The tool magazine is based on the operational principle of random selection and has a capacity for 24 tools, or as an option 48 tools or another 72 tools. The positioning tolerances of the horizontal machining center correspond to those of the CNC universal milling and boring machine. A further step was the equipping of the horizontal machining center with a pallet – changing system and an automatic tool changer.

Small and Average – sized Batches

The standard machining center completely equipped as just described , with its enclosed work area and the capability of loading and unloading pallets outside the work area while another is being machined, is even better suited for batch production than the universal milling and boring machine previously described.

Through his concept for one – shift operation the breaks and distributing time of the operator can be incorporated into the main machining time. The most salient features of the machining center are to be found in the way in which it at any time can quickly accomodate any changes made in batch numbers and , or in the types of workpieces during the manufacture of avarage – sized batches.

Large Batches (Substitute for less flexible production lines)

As a result of growing product variety, constant product improvements and the design changes involved here, as well as being a result of dynamic market developments, the actual manufacture of large batches is steadily dropping off.

Special production lines for manufacturing large batches, which in the past were regularly invested in, are now being replaced more and more by flexible manufacturing cells.

Large – batch manufacturing in conjunction with flexible manufacturing cells is done by the multi – combination of a machining center with a pallet – storage system, for workpieces weighhing upwards of 30 kg or a machining center with robot feed.

The machining center with pallet storing consists of a machining center and tool – changing system compliment by a pallet storage – system. For this purpose a pallet– storing system was designed and constructed, in which a large number of workpieces identical or varied are automatically fed to the machining center. In the case of the machining center with robot feed the basis is likewise provided by the machining center and pallet – changing system with the addition of a robot for loading the pallets.

The robot removes the workpieces from a storage rack and places them in the fixture. After machining has been completed it takes the workpiece back out of the fixture and places it in the storage rack.

The entire workpiece – changing process can be done during main machining time. A problem involved in robot handling is frequently caused by the variety and complexity of the various workpieces included in the large batches. This would make it necessary for the robot to have a large number of various mechanical hands with a range of different grasping strengths.

The solution presented itself as the possibility of glueing transporting bolts onto each workpiece, which can be grasped by a simple mechanical hand. In addition the transporting bolt has an advantage in that it contains a preorientation and thus enables the robot to place each workpiece correctly back into the storage rack without any complications.

The glueing of the transporting bolt onto the part can be done in an automatic glueing fixture. This process takes about 10 – 20 seconds.

The parts thus equipped with transporting bolts are then placed in a storage rack by the robot and are now available for machining. Of course this type of manufacturing concept requires an adequate monitoring system such as that described above.

Concluding Remarks

The current trend in the machine tool industry is basically towards standard machines equipped with standard units, since by way of batch production, they can be manufactured more cheaply.

Thanks to the productivity of modern controls, machining problems can be solved today by special individually – designed software.

In the years to come, the general tendency will continue at an increased rate, since continued growth in the productivity of micro – processors can be expected.

Universal milling and boring machine

The CNC machine feature full contouring control in up to 4 axes. D.C. feed drives and recirculating ball screws are standard equipment. A convenient and easily operated CRT screen contouring control for up to 4 axes is available.

Tool – Changing Equipment for the Universal machine

Tool Magazine

Flexible manufacturing cell

As the central modular component and with the right peripheral equipment, the machining center can be modified into an flexible manufacturing cell. Automatic self – supervised operation is made possible by the pallet pool, which enables job changes within a single setup, and the large capacity tool changer for automatic tool presentation when they are combined with monitoring and controlling equipment. This unit enables uninterrupted production during breaks as well as low – manned operation during regular shifts. This manufacturing cell can be used economically both in the large – scale production of a certain single part and in manufacturing various discrete parts. A memory – programmed control system supervises the pallet pool, which operates according to the principle of random selection. For organization and access to all data of the system it is also convenient to have a cell computer with the appropriate software. However, the cell is also completely functionable without such a computer. The computer assumes the task of organizing and monitoring the entire flexible manufacturing cell. In addition to the supervisory control of the system components such as machining center, pallet pool and tool magazine, the cell computer also maintains contact with peripheral equipment such as printer, tool – presetting device, macro printer and reader, as well as with the central computer.

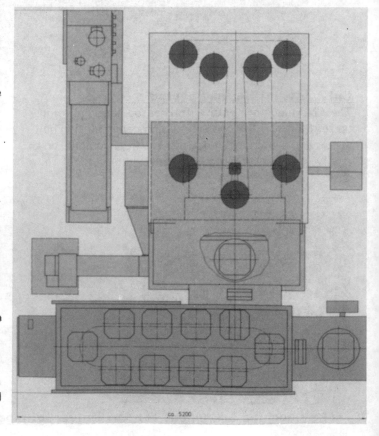

98

Changes in CNC Turning Technology

Anton W. Berkemeier, Singapore

It is now 112 years since a lathe was equipped with a camshaft and thus the first single-spindle automatic was introduced. For a full century this technology has contributed to a continuous progress in the production of turned parts in large quantities. The technique has been continuously refined in many details, was soon developed into multi-spindle automatics and transferred into other machining processes. Together with the development of the single-spindle automatic and later the multi-spindle automatic went an admirable development in cutting tool technology. When flowforming of metal parts and plunge grinding was introduced the enthusiasts of these new developments predicted the death of the automatics: their method would soon replace the turning of parts produced in large quantities and the small quantities would be left for CNC lathes. This prediction was wrong.

CNC Development in Lathes

At the time when the first CNC lathes entered the market for the production of components usually machined from bar stock, the program controlled automatics had reached their largest market share. The capability of these program controlled automatics with their extensive variety of accessories had reserved a particular area of application for these machines: the medium quantity production of parts that was uneconomical for cam controlled automatics. With shorter setting times they were comparable in production output but their higher cost did not allow for their use in large quantity production. Their only disadvantage was that they still required formtools.

The first CNC automatics capable of playing a competitive roll in this field were introduced in 1975. These machines with 65mm bar and 200mm chuck capacity had with a 22kW workspindle drive the necessary power for fast metal removal and on the other side due to an uncompromised design the ability of turning tightest tolerances with very high consistancy. Their application therefore soon shifted in approximately 30% of all cases towards the finish turning of components which previously required subsequent grinding. It was in these cases that for the first time CNC automatics were used for larger quantity production. By then many CNC lathe manufacturers had also realised the importance of the number of tools available in the machining process to allow for the maximum number of simple, single point cutting tools.

The next step in development was the introduction of tool drives for off-centre drilling and milling operations together with either a workspindle brake or a spindle indexing attachment. Soon the manufacturers of CNC controls helped us further with the development of the controlled movement of the workspindle, the so-called C-axis. This feature alone was relatively cost intensive, especially on larger machines where a wormwheel drive is needed for a higher torque on the spindle. With only a small number of revolving tools available the justification was usually only given by the rare odd component based on the calculation of setting, fixture and machine cost for subsequent operations while the CNC lathe with C-axis and revolving tools could finish the component in one operation.

CNC technology today

The introduction of the c-axis can be regarded as the last important development to today's CNC turning technology which had an important influence on further improvements and surrounding developments. And yet, many manufacturers still do not offer this function. This can only be understood as a reflect immage of a slow market response to important changes in manufacturing technology. The possibilities offered by modern CNC technology today requires a very different approach on the manufacturing and planning level. It might take some more time to make people understand that yesterday's methods are so quickly being overruled by these new developments.

The simplicity of programming standard workpieces has already given the CNC lathe or CNC automatic a wide field of application, also for larger batch quantities. This cannot only be seen under the aspect that people capable of calculating and designing cams and capable of setting the conventional machines have become scarce. In important factor also lies in the unreached advantage of CNC controls in guiding a single point cutting tool along a contour at very high speed, very accurately and by this in the drastic reduction of the cost for cutting tools per production output. Here lies one of the very important advantages of CNC turning.

An important and interesting field of application for CNC lathes has developed out of the fact that machines can be designed to make full use of the smallest CNC control increment: the capability of controlling the cuttingtool movement with the accuracy of one Micron (0.001mm). This is the field in which our company mainly operates. If the machine has the capability of repeating a defined contour position within this micron on the component under production condition then it should be used as the new means for finishturning many components and thereby avoiding subsequent grinding operations. This conclusion, however, is only one step and it cannot stand alone. For the full implementation of CNC turning as the general turning method, further steps are required.

With relatively conventional design methods the above described criteria were achieved in machines for components larger than 50mm diameter: high accuracy, c-axis and revolving tools opened a wide range of application. We now face the question of a higher cost efficiency. The demand for a second independant tool carrier is therefore very strong, especially for larger quantity production. In the range of components under 50mm diameter this demand is even more eminent than on larger machines which will quickly be explained by a simple calculation: The influence, a bar of 20 or 30mm diameter and 3 Metre length revolving at 7.000 r.p.m. has on the accuracy of the product should at this point be neglected. Small components have small dimensions in faces and length.

Roughing for example a 15mm diameter over a length of 20mm and then facing it up to the bar diameter of 20mm is a total travel of the cutting tool of 23.5mm including the approach at the beginning and the clearing of the tool at the end of the cut. With a feedrate of 0.15mm/rev. this cut will only take 1.35 seconds at 7.000r.p.m. The efficiency of such operations becomes highly disputable when further looking at the short times required for performing small recesses or the finish turning of short diameters in relation to the out-of-cut time for turret indexing etc. On many machines indexing takes more than 1 or 2 seconds. The question of efficient machine utilisation is therefore very vital on small CNC machines.

As the example shows the metalcutting or productive time in such processes will easily correspond to less than 50% of the total machining time. Smaller NC machines must therefore have the shortest possible idle times, a task not easy to achieve. Even with highest rapid slide movements and indexing times of 0.3 or 0.5 sec. the proportion of unproductive time in the machining time will still remain relatively high. The trend in small CNC machines will therefore be very much towards 4-axes machines with two simultaneously operating turrets. These solutions permit to overlap the indexing and idle movements of one turret by the operation of the other. 4-axes machines have the further advantage of increasing the total number of tools available which is critical for small components. A strict rule for CNC turning is simple tooling for each operation. This rule is often not easy to observe when bore diameters get small. The experience shows that most turret positions will be required for internally operating tools when machining small components.

With 4 axes, c-axis, revolving tools and the possibility to finish turne tight tolerances we have the prerequisit for covering almost all turning problems. And here the new way of thinking in CNC turning starts: as shown before it is today no more a question of whether to use a CNC automatic or not when the criteria is the high accuracy of a component. Diameter tolerances of IT6 can be turned on most materials under production condition, also when IT6 means no more than a tolerance range of 0.005mm on small components. Efficiency is guaranteed through lowest tooling cost, short machining times and by the reduction of subsequent operations. Here in particular the c-axis and the revolving tools play the important role. The machine shown in **Fig.1** offers a maximum of 10 revolving tools with 2,2kW DC drives on both turrets.

With this large number of revolving tools these machines present a combination of manufacturing facility that was unknown before. Highly accurate turning operations can be combined with extensive off-centre milling and drilling operations which before had to be taken to small machining centres. Typical components are shown in **Fig.2, 3 and 4**. They previously presented the manufacturing side with subsequent clamping problems which were almost impossible to solve. The flud of such components coming to these CNC turning centres now have also lead to a further development: the synchronised spindle in turret 1 and the additional rear-end drilling station. Both can be seen in Fig.1 on the main turret respectively to the right of it. This spindle uses the same 2.2kW DC drive as the revolving tools in turret 1 and can be synchronised with the workspindle revolution to 4.750r.p.m. During the part-off operation from

turret 2 the component is clamped by this spindle and will then be indexed to the rear-end drilling station. While the machining of the next component will be started at the main workspindle by turret 2, the previously machined component will undergo further operations on the part-off side. It must be mentioned that the synchronised spindle can be indexed every 90 degrees and the rear-end drilling station can also be equipped with revolving tools. This offeres additional off-centre milling and drilling operations on the part-off side.

Trends

The title CNC Turning Centre for these machines is more than justified. As a turning centre they must have the the provision for automatic loading and unloading of components from standardised pallets and for automatic pallet handling, **Fig.5**. After encouraging first attempts, practicable solutions are also on their way for automatic tool changing and the exchange of workholding elements. In-process and post-process measuring are still relativly cost intensive options but commonly in use where justifiable by the high cost of precision components.

A much different side of CNC turning is the field of multi spindle automatics. Our first full scale CNC 6-spindle automatic exists and has up to 32 CNC axes,

Fig.6. It is again a straight forward CNC design where all tool movements are CNC controlled. CNC turning today stands for single point contour turning and therefore no formtools are required on this machine. To achieve this the upper four spindle positions operate with two cross slide systems each. The consequent CNC thinking has resulted in new design features for fast changing of material, feed and clamping collets. The 10 cross slides offer the same possibility for a standardised tooling as is available on single spindle CNC machines. Fast exchange of presett tooling and high spindle speeds for the use of modern coated carbide tools are also standard features.

CNC turning today covers the full range of application, from simple operations to the combination of highly accurate turning plus contour milling and drilling. CNC turning today presents a completly new understanding of the word turning. Who ever has to produce turned parts today faces a highly competitive market. A world-wide overcapacity in turning machines forces to cuts in manufacturing cost for survival. Cost reduction respectively an increase in productivity can today only be achieved by a reduction in material flow time, i.e. by avoiding operations on further machines.

Fig.1:   5-axes CNC Turning Centre with synchronised spindle
          and rear end drilling station

Fig.2, 3 and 4:
          Components requiring C-axis and revolving tools

Fig.5:   CNC Turning Centre with integrated computerised
          workpiece and pallet handling system

Fig.6:   Maching area of CNC Multispindle Automatic

Fig 1

Fig 2

Fig 3

Fig 4

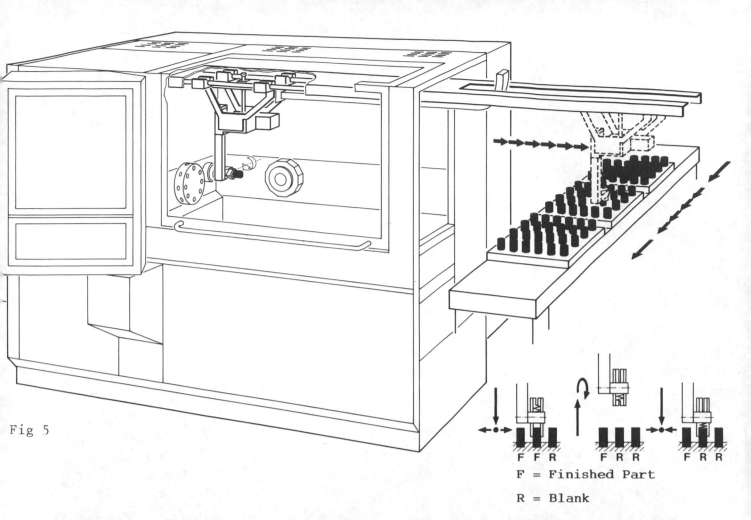

Fig 5

F = Finished Part

R = Blank

Fig 6

MODERN TECHNIQUES IN GRINDING WHEEL FORMING AND DRESSING

J. STEPTOE (DIRECTOR) PRECISION GRINDING LTD. ENGLAND.

The method of forming grinding wheels is usually determined by the nature of the workpiece geometry, production volume and standards of accuracy required.

A variety of techniques are in current use employing such standard workshop equipment as sine bars and slip gauges in conjunction with block-mounted diamond tools(Fig 1) that allow of simple, accurate dressing of a limited range of angles. Radii can be dressed with table-mounted attachments that require the use of gauge blocks and dial gauges for initial setting and checking of diamond wear.

As the geometry of the workpiece form progresses to the inclusion of blended tangent angles and radii (Fig 2), projectors, microscopes and similar aids are required for checking the wheel form. This imposes demand on time in operation and skill in interpreting the measurements. Alternatively, these more complex forms may be dressed by the employment of wheelhead-mounted dressers that incorporate co-ordinate micrometer slides and a 5:1 or 10:1 ratio pantograph linked to a dressing arm carrying a lapped diamond or dressing tool precisely matched to the accuracy of a 5:1 or 10:1 stylus. Diamond wear cannot be monitored during the dressing operation.

With this equipment, the manufacture of tooling to suit a particular workpiece involves calculation of the template dimensions at the appropriate ratio and, according to the workpiece dimensional tolerance, production by milling, jig boring and/or grinding (Fig 3) with subsequent inspection to verify accuracy. For this last procedure, further time may be sometimes added by the preparation of a magnified projection layout before the unit is ready for operation. The tooling, in short, can frequently take as much time in its manufacture as that to produce a total batch of low-volume components which, incidentally, will still require off-machine inspection.

Although wheel forming by numerical control represents the highest capital investment in this field, it still lacks the capability of monitoring dressing tool wear, vital to the maintenance of consistent accuracy. Furthermore, when costing-out workpieces produced by this equipment, allowance must be made for re-working or replacement of precision-lapped dressing tools. Additionally, the costs of machine down-time during the usual tape control try-out and inspection procedures before committal to actual production should be assessed in evaluating this method.

In considering these various techniques, it becomes apparent that each is susceptible to one or more strictures upon which their function is dependent. These may include need of the following auxiliary facilities: Setting and measuring devices - Specially lapped diamonds or dressing tools - Form templates - Prepared projection over lays - A means of measuring wheel wear - Tape programming and try-out - Off-machine job inspection.

A system of wheel dressing by which these factors can be minimised or, better still, eliminated, merits close examination and the adaptation of today's miniaturised microprocessor technology to this area of workshop practise has, happily, enabled this to be achieved.

OPTIDRESS (Fig 4) is a unit that adapts for fitting to a wide variety of surface grinding machines of British and overseas manufacture. It is totally self-contained and functions independently of the ancillaries already described. A logical successor to its well known predecessor that utilised an optical and mechanical micrometer measurement combination for dressing radii and tangent angles, the new OPTIDRESS E retains the optical system but employs micro-electronics for setting dressing tool datums and measuring each movement made in forming the wheel.

The new instrument's development was not prompted merely for the sake of change but for two important reasons. Firstly, improved accuracy. The digital read-out system has the capability of resolving much finer measurement than that obtainable from a micrometer scale and offers the advantage of displaying true readings unaffected by leadscrew backlash or pitch errors. A high degree of accuracy in wheel forming is absolutely essential in meeting the very tight tolerances laid down for form ground tooling used to produce small, precisely mated parts in the electronics industry, as a typical example.

A second, equally valid reason for adopting digitised measurement was to maximise simplicity of operation and enable people of only moderate skills to achieve these degrees of accuracy. When dressing a complicated form on a grinding wheel, it is comparatively easy for the machine operator, skilled or not, either to confuse one micrometer reading with another or miscalculate the required number of turns of a dial. The OPTIDRESS FACTSTOR read-out is positive, easily understood and virtually foolproof.

The OPTIDRESS MICROSCOPE and built-in graduated reticle serves as a means both of centering the dressing tool for establishing the datums in the FACTSTOR memory and for observation of the magnified profile of the grinding wheel during the forming operation We have touched more than once upon the importance of ascertaining the amount by which the dressing tool wears. This applies equally to the wheel during the grinding cycle and occurs regardless of the dressing method employed. The OPTIDRESS microscope allows both these conditions to be immediately observed and compensated for. The breakdown in the form in this picture is clear and unmistakable and indicates 'REDRESS' (Fig 5)

The microscope tube acts as a pivot around which rotates a radius arm carrying the radius and tangent slides and dressing arm (Fig 6). By this arrangement, the radial and tangential movements are made in relation to the optical centre from which the datums are set, the degree of rotation being governed by a series of stops that permit pre-setting of up to three pairs of angles.

Positioning of the complete assembly in relation to the grinding wheel is effected through co-ordinate longitudinal and cross slide movements.

The unique feature of the FACTSTOR digital read-out unit (Fig 7) is that, once the optical datum settings are recorded in its memory, subsequent dressing movements in each of four axes is displayed as a single read-out measurement of that axis and that axis alone. As soon as a movement in another axis is executed, the reading changes correspondingly. At all times, however, measurements of previous settings can be recalled at the touch of the appropriate axis switch, a facility that allows for clear, uncomplicated readings in metric or inch measurement to 0,005mm or 0.0002in of the linear movements. Angular reading is to within 1 minute of arc. As mentioned previously it is not practical to resolve these fine linear increments on a micrometer scale nor angular increments this small on a vernier.

The form being dressed on the grinding wheel is, of course, only as true to the FACTSTOR readings as dressing tool wear permits but, with the OPTIDRESS, it is a very simple matter to control this factor by re-setting the tool to the zero datum, checking the amount of wear against the microscope reticle graduations and, if unacceptable, re-setting to the reticle cross-line.

This may well be an appropriate point at which to mention the nature of dressing tools themselves. (Fig 8). Traditionally, natural diamonds have been used in this respect and their cost and comparatively brief life-span have added significantly to the busy toolroom or trade shop budget. As grinding specialists, my Company has devoted considerable research to this subject and, as a result, have developed a radically new style of dressing tool for use with the OPTIDRESS, known as POLY-TIP. POLY-TIP is triangular in shape and is mounted in a special toolholder at the wheel centre-line with a flat top rake so that its tip cuts the grit rather in the manner of a lathe tool As each tip becomes worn, POLY-TIP is rotated to a new position and, finally, turned over for a repeat of the procedure. By this means, six dressing points can be utilised, the life of each one of which, we find from our tests, equalling that of a natural diamond. Quite apart from the major economies offered by this tool, its clean cutting action produces an open, non-impacted grit structure in the wheel, excellent heat dissipation and an avoidance of the problems to which brazed shank-mounted diamonds are liable.

We have not, so far in this discussion, turned to the subject of monitoring the grinding operation itself. (Fig 9) To verify the depth and position of the ground form in the workpiece, often by more than one inspection procedure in a separate location from that of the machine, can be wasteful, time consuming and costly. In addition, it is by no means easy, in some cases, to re-locate the workpiece precisely in the same position on the chuck or fixture with the result that, in the final stage of the operation, the job can be scrapped.

The PROJECTORSCOPE (fig 10) is designed to complement the OPTIDRESS or other wheel forming attachments so that these problems may be eliminated. PROJECTORSCOPE is carried upon a precision micrometer horizontal and vertical slide arrangement mounted at one end of the grinding machine table. At the other, a tungsten-halogen illuminator projects a light beam so that the magnified profiles of wheel and workpiece are projected on to a ten-inch square screen and compared with a selection from a range of turret-mounted angular, radial and cross-line reticles built into the optical system. Alternatively, a prepared overlay drawing may be used when grinding more complex forms or those that extend beyond the area of the screen. In the latter case, the desired profile is broken down and drawn in overlapping sections.

The PROJECTORSCOPE horizontal and vertical slides give the operator complete assurance of correct positioning of the wheel in relation to the workpiece, irrespective of the condition of the machine leadscrews. Having used the micrometers to set the workpiece datums on the screen, he positions the magnified wheel profile in the required relationship and commences the grinding operation. By reference to the reticles or screen overlay, he is able to confirm that the depth of the form is to drawing, that the form is correct and that the job is completed.

CONCLUSION

To meet the growing demand for accurate form-ground parts, it is essential to combine the optimum use of skilled labour with equipment that allows for on-machine inspection and minimal down-time, factors that may represent the difference between profit and loss when working at highly competitive rates.

Further cost savings can often be effected by a review of the toolroom planning method of precision tool production. The use of modern wheel dressing equipment capable of forming complex shapes in minutes without auxiliary aids enables the form to be ground directly into the hardened blank. This is often vastly preferable and very much more economical than conventional machining of the part in the soft state prior to heat treatment. At this stage, problems can arise from incorrectly machined grinding allowance distribution, sometimes compounded further by distortion in heat treatment. Each machining process eliminated makes for quicker delivery of the finished part.

My own Company's policy lies in the continuing development of easy to use, foolproof equipment designed to minimise the pressures on supervisory staff and to make life easier for the operator in achieving today's demands for high precision tooling.

Low Cost Automation (LCA)
Through Special Purpose Machines (SPM)
with the Help of Pneumatics

Dipl.-Ing. Kurt Stoll
FESTO KG
Esslingen, Federal Republic of Germany

Low Cost Automation is an intermediate technology appropriate to all industries for improving productivity. It is used to modify existing machines by adding readily available simple devices. Very often Special Purpose Machines play a key role: with one mostly imported central part and many locally manufactured peripheral parts they cover a wide range of the industrial handling market. Together with Pneumatics with its simple and flexible standard elements and its easy-to-learn technology small and medium sized production can effectively be mechanised to gain more safety, better quality and higher productivity.

1   INTRODUCTION

How is a pair of scissors manufactured ?
Metal must be machined with knowledge and experience to produce, with effort, a pair of scissors (fig. 1).

This path from the metal to the scissors - or from the material to the product - is divided into steps. These steps represent a manufacturing process. Central to the manufacturing process is the human being, who ensures that the individual steps are carried out in the correct sequence. The human being controls the manufacturing process.

If we relieve the human being of more and more control tasks through the application of technical aids - the process will finally run itself, i.e. automatically.

What does "Low Cost Automation" mean ?

It means the liberation of man from dangerous and/or repetitive (often stupid) manual activities, with an investment at low cost in a simple and easy-to-learn technology.

What are Special Purpose Machines ?

With an imported center part and all other locally manufactured peripheral parts these machines cover several steps in the field of industrial handling operations.

What help can be given by Pneumatics ?

With its simple technology, flexible standard elements and easy-to-work-with energy, Pneumatics can be applied in all handling operations, such as (fig. 2 + 3):

- bunkering
- magazining
- transporting
- positioning
- clamping
- turning
- feeding
- linear driving
- rotary indexing working
- processing
- ejecting

In which industries can we find applications of Pneumatics for Low Cost Automation ?

As mentioned above, pneumatics is a number one help in all handling operations. That is why we find pneumatics in almost all industries. Focus is on the metal industry, but the woodworking industry, especially, very often requires pneumatic devices. The textile industry, mining industry and also agriculture are areas of various applications.

2   CHARACTERISTICS OF LOW COST AUTOMATION

Low Cost Automation is an intermediate technology appropriate to all industries for improving productivity. It is used to modify existing machines by adding readily available simple and flexible, standard mechanical, pneumatic, hydraulic, electric and

112

combined devices at relatively low additional cost to substantially increase production rate without change in manpower (fig 4):

- The output of existing machines can be increased by adding simple devices.
- The devices can be fixed by do-it-yourself.
- Only a minimum designing work is necessary.
- The devices are readily available and re-usable for different purposes.
- Modifications can be carried out step by step.
  That helps to go from old to new.
- Short pay-back period of investment.
- Applicable for small and medium batch production (50 to 5000 pieces).

3 CHARACTERISTICS OF SPECIAL PURPOSE MACHINES

Special Purpose Machines play a key role in the field of automation: between the highly sophisticated industrial robot and the old manually operated work place they help to improve both - the human labour efficiency and the mechanical work output.

Depending on the actual function and situation some parts of these special purpose machines must be imported, but some can be manufactured locally with the result of saving money and foreign currency (fig. 5 + 6):

- The application of standard modules allows a great flexibility.
- There is only a minimum time for resetting or retooling necessary,
- Several - mostly peripheral - parts can be manufactured locally, only a
  few parts - so called center parts - have to be imported.
- With this division of labour the costs can be reduced by approximately 50 %.

4 CHARACTERISTICS OF PNEUMATICS

Pneumatics is a low cost technology with compressed air as the energy source. Simple standard elements at low cost from the shelf, together with an easy-to-learn technology, guarantee an economic and quick application without a university degree (fig. 7):

- Air is available without limitation.
- Energy in the form of compressed air can easily be transported without
  the need of return.
- The energy is storable in any quantity.

113

·   Compressed air is safe from explosion hazard and fire risk.

-   The energy is clean and will not pollute the environment.

-   Pneumatic working elements are simple and inexpensive.

-   Floorshop workers can easily learn this technology.

Special attention must be paid to the following technical features of pneumatics (fig. 8):

Five times infinitely variable:

- Force	1	... 2500	kp
	10	... 25000	N
- Speed	50	... 2000	mm/sec
- Time	0.3	... 300	sec
- Stroke	1	... 10000	mm
- Pressure	0.1	... 10	bar

5   EXAMPLES OF LOW COST AUTOMATION THROUGH SPECIAL PURPOSE MACHINES WITH THE HELP OF PNEUMATICS

Due to the great number of similarities between the characteristics of low cost automation, special purpose machines and pneumatics there are many successful examples in various industries (fig. 9 ... 18):

-   metal industry

-   wood industry

-   textile industry

-   packaging industry

-   agriculture

-   food industry

6   SUMMARY

Low Cost Automation through Special Purpose Machines with the help of Pneumatics is of great importance to all industries. With relatively small investment costs it provides great possibilities of creating work for the unemployed and raising productivity (fig. 19).

Pneumatics is the ideal technology for the successful realisation of Low Cost Automation in order to achieve more safety, better quality and higher productivity.

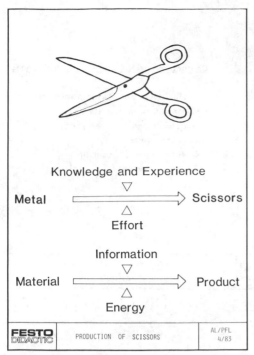

Knowledge and Experience

Metal ———————▽————————> Scissors
          △
        Effort

Information

Material ——————▽————————> Product
          △
        Energy

FESTO DIDACTIC | PRODUCTION OF SCISSORS | AL/PFL 4/83

Fig.1.

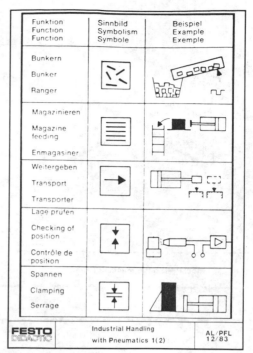

Funktion Function Function	Sinnbild Symbolism Symbole	Beispiel Example Exemple
Bunkern Bunker Ranger		
Magazinieren Magazine feeding Enmagasiner		
Weitergeben Transport Transporter		
Lage prüfen Checking of position Contrôle de positon		
Spannen Clamping Serrage		

FESTO DIDACTIC | Industrial Handling with Pneumatics 1(2) | AL/PFL 12/83

Fig.2.

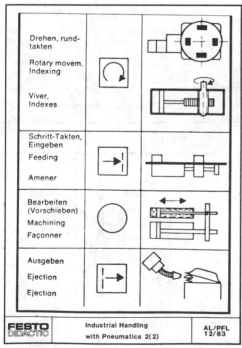

Drehen, rund-takten Rotary movem. Indexing Viver, Indexes		
Schritt-Takten, Eingeben Feeding Amener		
Bearbeiten (Vorschieben) Machining Façonner		
Ausgeben Ejection Ejection		

FESTO DIDACTIC | Industrial Handling with Pneumatics 2(2) | AL/PFL 12/83

Fig.3.

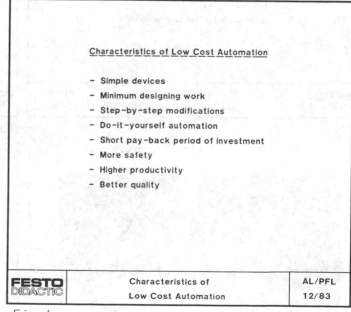

Characteristics of Low Cost Automation

- Simple devices
- Minimum designing work
- Step-by-step modifications
- Do-it-yourself automation
- Short pay-back period of investment
- More safety
- Higher productivity
- Better quality

FESTO DIDACTIC | Characteristics of Low Cost Automation | AL/PFL 12/83

Fig.4.

115

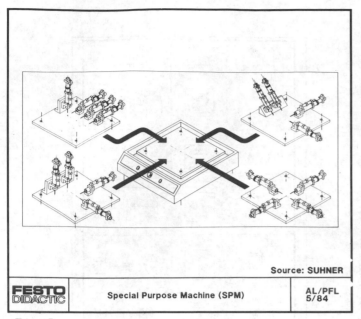

Source: SUHNER

| FESTO DIDACTIC | Special Purpose Machine (SPM) | AL/PFL 5/84 |

Fig.5.

Characteristics of Special Purpose Machines

- Standard modules allow great flexibility
- Minimum time for resetting
- More local manufacture
- Less import parts
- Cost reduction of about 50 %
- Saving money and foreign currency

| FESTO DIDACTIC | Characteristics of Special Purpose Machines | AL/PFL 5/84 |

Fig.6.

Characteristics of Pneumatics

- Simple technology
- Flexible standard elements
- Easy – to – work – with energy
- Air unlimited available
- Energy storable in any quantity
- Safe from explosion hazard
- Clean energy
- Suitable for low cost automation

| FESTO DIDACTIC | Characteristics of Pneumatics | AL/PFL 12/83 |

Fig.7.

Technical Features of Pneumatics

Five times infinitely variable:

– Force	1 ... 2500 kp
	10 ... 25000 N
– Speed	50 ... 2000 mm/sec
– Time	0,3 ... 300 sec
– Stroke	1 ... 10 000 mm
– Pressure	0,1 ... 10 bar

| FESTO DIDACTIC | Technical Features of Pneumatics | AL/PFL 12/83 |

Fig.8.

| FESTO | SPM for manifold mounting | AL/PFL 6/84 |

Fig.9.

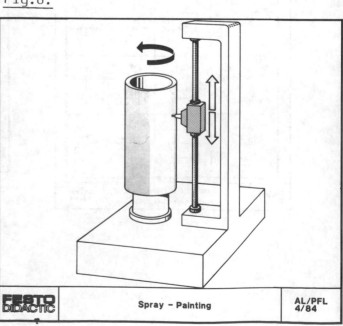

| FESTO DIDACTIC | Spray – Painting | AL/PFL 4/84 |

Fig.10.

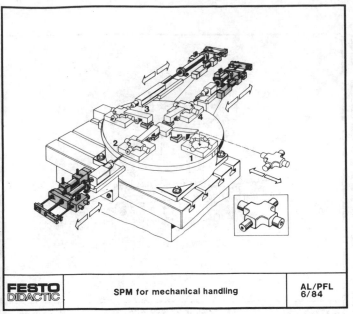

| FESTO DIDACTIC | SPM for mechanical handling | AL/PFL 6/84 |

Fig.11.

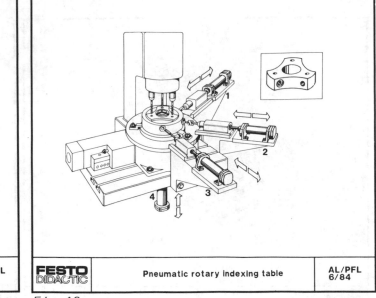

| FESTO DIDACTIC | Pneumatic rotary indexing table | AL/PFL 6/84 |

Fig.12.

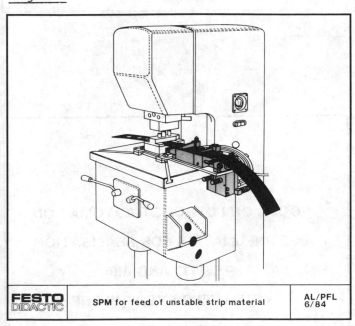

| FESTO DIDACTIC | SPM for feed of unstable strip material | AL/PFL 6/84 |

Fig.13.

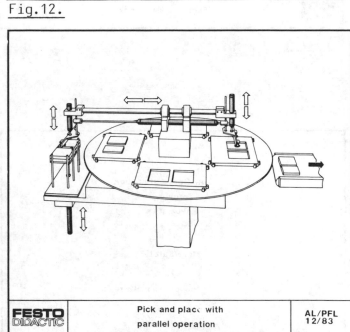

| FESTO DIDACTIC | Pick and place with parallel operation | AL/PFL 12/83 |

Fig.14.

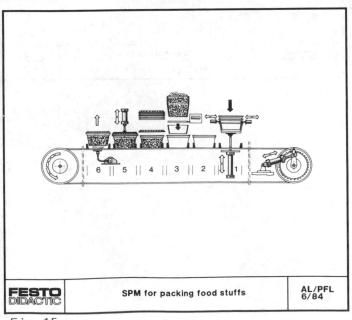

| FESTO DIDACTIC | SPM for packing food stuffs | AL/PFL 6/84 |

Fig.15.

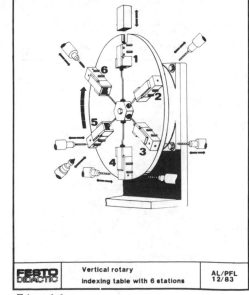

| FESTO DIDACTIC | Vertical rotary indexing table with 6 stations | AL/PFL 12/83 |

Fig.16.

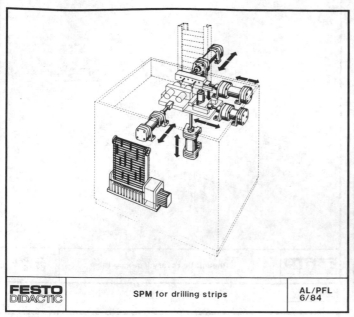

| **FESTO** DIDACTIC | SPM for drilling strips | AL/PFL 6/84 |

Fig.17.

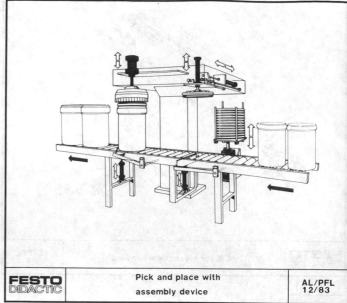

| **FESTO** DIDACTIC | Pick and place with assembly device | AL/PFL 12/83 |

Fig.18.

Low Cost Automation
Through
Pneumatics

- More Safety

- Higher Productivity

- Better Quality

| **FESTO** DIDACTIC | Results of Low Cost Automation | AL/PFL 12/83 |

Fig.19.

FOCUS ON LOW COST AUTOMATION

USE PNEUMATICS FOR REALISATION

FESTO AND AIR

YOU'LL FIND EVERYWHERE

| **FESTO** DIDACTIC | Low Cost Automation Through Pneumatics | AL/PFL 12/83 |

Fig.20.

USING CAM TO OPTIMISE THE BENEFITS OF AUTOMATED BATCH PRODUCTION

By Alan D. Murphy, Cambridge Consultants Ltd, UK.

SUMMARY

Cost effective automated batch production is the ultimate aim of most manufacturing industries. With growing consumer sophistication worldwide the mass-production industries are having to provide a series of customised products in smaller runs, more responsively and at lower cost. Many large organisations have taken the costly FMS route for component production, but the majority of manufacturers are medium sized organisations who cannot expose themselves to such high risk projects. This paper explores an alternative route explaining how computer aids can optimise the benefits of automated batch production, using case histories as practical examples of the approach.

1. INTRODUCTION

The majority of world manufacturing in the engineering and allied industries is batch by nature. The sizes of batch vary enormously from as little as 1 in custom machinery, through 10 for specialist equipment, 25 for aero products, 100's for trucks, 1000's for specialist cars, tens of thousands for consumer electronic products. Whatever the actual batch size the problems of control are usually pre-eminent. In recent years the maturing of computer based systems for materials and resource planning, capacity planning, machine control etc has given operations management a set of tools that can have a tremendous effect on company performance. As these tools become increasingly more integrated (CIM) the effect will be to further optimise effectiveness by reducing inventory, lead times and indirect labour costs to a minimum. There are companies such as Volvo, Black and Decker, Toyota, Sony etc who compete on the basis of operational excellence which is equivalent to saying that they put CIM at the heart of their business strategy. Most of the companies that publicise this approach are household names with major financial resources, however many medium sized manufacturers are making headway in this area with limited investment budgets.

This paper illustrates the advantages of using computer aided manufacturing (CAM) techniques by a series of case studies in the following areas:

* Electronic component production
* Printed circuit board (PCB) assembly and test
* Yarn packing and warehousing
* Meat slicing and packing

2. CASE STUDY 1

In 1978 a Johnson Matthey subsidiary that manufactured platinum film resistors in small batch quantities requested CCL to look at the feasibility of automating the process. It was felt that automation would improve the performance of the product (graded to B.S. procedures) thus allowing a higher component price to be available. Hence the stimulus to automate came from a business strategy to increase volume and increase contribution from each component. A joint task force between CCL and the client was set up to assess the feasibility as both the product and the process required development in parallel.

At that early stage of the project the team agreed that the following processes were required to produce high grade product:

* Coat the ceramic substrate with platinum ink by a thick film dipping process.
* Dry the coating.
* Coat a second time.
* Dry the second coat.
* Transfer dry product to kiln furniture for firing.
* Transfer fired product from kiln furniture to laser pattern cutting.
* Anneal product.
* Trim pattern to final resistance value.
* Weld terminations to product and add solder reinforcement.
* Coat product with glass and dry.
* Fire glass coating.
* Measure product resistance at 0°C and 100°C and grade to B.S. specification.

The component was being manufactured by labour intensive means with relatively low yield of grade 1 and 2 product during this investigation, however the background knowledge and the opportunity of trying various experiments was extremely useful.

The initial study yielded the following requirements for automated equipment:

* Automatic two stage dipping and drying machine.
* Automated transfer of dry (but still green) substrate to kiln furniture.
* Automated laser pattern cutting and testing.
* Automatic transfer of cut substrate to kiln furniture.
* Automated laser trimming and testing.
* Automatic welding of terminations and solder applications.
* Automatic spraying and drying of glass coating.
* Automated measurement of grading parameters and delivery to graded containers.

In essence the process broke down into five automated machines with inter-process firing operations using tunnel kilns. Testing at each stage was imperative to monitor process changes as they progressed, these early parameters checks were critical to final product grade. It was agreed that product would be made in daily batches of 2,000 on single shift giving approximately 500,000 p.a. capability.

The technical problems that had to be overcome are too numerous to mention here, but some that were critical to success were as follows:

120

* Final shape of ceramic substrate to enable automatic orientation for feeding purposes.
* Consistency of the dipping process.
* Design of kiln furniture to allow automatic entry of green coated substrate without causing surface damage.
* Computer control of the laser pattern cutting and trimming machine.
* Welding of platinum or nickel terminations.
* Design of holding jigs for the glass spraying machine.
* Design of four terminal measurement connectors for automated test machine.
* Computer control of automated grading with trend analysis.

The client continued to refine the process as CCL developed the equipment and installation commenced late in 1980. Trial running started in Spring 1981 and detailed changes were made as required. However the basic substrate design and associated machines remained the same. Some refinement occurred e.g. the second dipping and drying process was found to be redundant, software was modified on the automated inspection machine to reduce the settling time and hence increase basic throughput.

The outcome was a controllable process that took only six operatives giving 80% high grade product. The investment was around £500K but the cost savings were approximately £1 per unit giving less than 2 year payback at predicted volumes.

In addition the computerised inspection machine has been utilised for other products in the range with similar physical characteristics.

3. CASE STUDY 2

There are an increasing number of medium sized electronics equipment manufacturers that are attempting to automate their PCB production facilities. As depicted in the first case study to optimise the returns of such investment requires a level of product redesign to be carried out in parallel. Our client in this area is a manufacturer of sound mixers for the professional market. He assembles and tests approximately 9,000 PCBs per month in batch quantities ranging from 1,000 off down to as small as 10 off.

By adhering to disciplined design guidelines as follows:

* Keep all components in the X and Y axes of the PCB.
* Ensure sufficient clearance between components.
* Keep centre distances constant.
* Ensure hole clearances are sufficient for ease of insertion.
* Provide convenient tooling holes for location.
* Keep a standard width of mother board to ease production handling.

It is possible to develop a step-by-step approach to flexible automated assembly using a mix of dedicated and robot insertion equipment.

The figures on the following pages illustrate the five steps towards a 95% level of automation currently reached in the Japanese TV industry.

* Figure 1 illustrates the basically manual approach to PCB assembly with a limited investment in ATE.

* Figure 2 illustrates the use of semi-automatic insertion equipment and the uprating of ATE.

* Figure 3 illustrates the flow line principle with automated handling between processes. This stage is typical of the more advanced UK manufacturers.

* Figure 4 illustrates the use of limited robotic devices for stick fed components and ATE board handling. This stage is typical of the large US manufacturers.

*   Figure 5 illustrates the ultimate approach with robot assembly replacing the majority of manual insertion operations. This is typical of the Japanese TV industry e.g. Sony, Panasonic and Toshiba, where 95% of all components are inserted automatically.

Our client is currenting going into stage 3 with planning for stages 4 and 5.

The investment requirements for semi-automated batch assembly of PCBs is as follows:

*   Automatic insertion equipment, the type being dependant upon the product.
*   Flow through conveyor, soldering and cleaning system.
*   Automatic test equipment (ATE) of the in-circuit variety with paperless rectification station.
*   Variety of manual assembly stations.

In this case the equipment list was as follows:

*   Fuji automated radial and axial inserter.
*   Electrovert conveyor, soldering and cleaning system.
*   Zehntel ICT with paperless repair station.
*   Amistar semi-automatic IC inserter.
*   4 manual assembly stations.

The overall investment is approximately £300,000 with potential annual savings of over £200,000 if used for the top ten PCBs which account for 80% of total volume.

The intention is to use this approach as the first investment towards a full blown paperless flexible manufacturing system (FMS) for PCB and module assembly.

The power of this approach is that the company only needs to assimilate a reasonable level of change at each stage and generates income quickly enough to help fund the proceeding stages.

4.   CASE STUDY 3

The problem of collating, marshalling, inspecting, labelling, palletising and warehousing cartons is common to many industries. If lines are dedicated to single high volume products e.g. baked beans, soups etc the flexibility and the use of batching is a minor requirement. But in many situations the level of throughput is not that high and plant has to be used for a number of product types at the same time. This was the problem of a CCL client in the chemical industry producing a high grade yarn product in Northern Ireland.

He was planning to increase spinning capacity by 50% by investing in highly automated plant and needed to improve his downstream packing and warehousing capability. His project engineers came to CCL with a proposed plant layout but little idea on how to effectively control the plant. Figure 6 depicts the layout of the system capable of handling 3 cartons, weighing around 50kg each, per minute. At any one time 8 different grades of product could be in the system albeit packed in identical cartons. Each carton had to be identified, logged into the system, placed in the correct accumulation conveyor, metered out when a pallet load of identical cartons were available, weighed, labelled, sealed and palletised. Completed pallets were transferred to the warehouse by fork lift truck.

The optimum control solution was a computer based system using DEC, LSI 11/23 hardware with twin floppy disc drives and standard input/output devices to interface with the plant sensors and actuators. The plant was under the control of a single operator who resided at the weighing and labelling station. Due to UK trade laws a human being has to accept the weight recorded by an electronic weighing device, hence acting as a final stage of quality control and consumer protection.

Some major technical highlights of the system are:

* Bar code reading using non-contacting optical means with a 2" depth of field. Practice has shown near 100% reliability.
* Computer control allows dynamic allocation of lanes to product.
* Overspill lanes for large pulses of product can be allocated, with overspill being taken out first to free the lane for further allocation.
* All plant fault conditions were transmitted to the computer system which then signalled the operator via the VDU with a clear message and the priority rating for rectification.
* User friendly operator interface to reduce training time to a matter of days.
* Recording of all carton and pallet data for automatic transfer to factory stock control system.

The major financial and operating benefits from the system were as follows:

* Only one operative for a system handling 50% more than previously.
* Due to the accumulation capability the packing plant can run for longer hours e.g. evening shift, with guarantee of clearance to warehouse next morning.
* On-line update of latest stock position to improve customer response times and help capacity planning.
* In-built diagnostics to maximise up-time.

The total investment was around £500,000 with the return coming from production, inventory and distribution control improvements rather than labour savings.

5. CASE STUDY 4

The use of computers to control machines is becoming more commonplace in the engineering industries e.g. machine tools, inspection equipment, robot systems etc, but in some industries computer control is still a rarity. CCL has developed CNC controllers for precision grinding machines in the past and has recently used this expertise to good effect in the food industry.

Thurne Engineering are a UK company specialising in the meat handling and slicing sector. They had developed a high speed meat slicing machine that had international market potential if they could supply a reliable, intelligent controller to drive the machine and interface with down stream equipment such as check weighers and packaging machines. Thurne's initial foray using electronic hardware techniques, was only partially successful, however using CCL's software driven CNC expertise and applying single board computer hardware has produced a successful product.

The machine comprises a slicing head, meat feed system, slice collating and handling arrangements all driven by separate motors. The machine is required to cut slices of a pre-selected thickness while conducting a number of concurrent operations plus interlock monitoring to ensure that operator safety is maintained.

Cut slices are collated on a slow moving conveyor which is accelerated to separate the programmed number of slices (typically 4 to 12) into packs. These packs are then weighed and packaged.

Technical highlights of the control are as follows:

* Performance monitoring is incorporated to log slice thickness and axis velocities.
* Slicing sequence and pack size can be chosen from a diverse range of programmable options.
* In built diagnostics ensure maximum up-time.
* Direct feedback from check weigher to maximise raw material utilisation.

The financial benefits of such a system are as follows:

* Accuracy and consistency of packed product with minimal give-away.
* Flexibility of pack specification and minimal lead time.
* Commissioning time reduced to one tenth of previous machines.

To summarise, the computer controlled machine gives the user a much more productive, reliable and flexible investment than previous lower technology equipment. This is analogous to the benefits derived from CNC machine tools in the engineering industry.

5. CONCLUSIONS

Computer aided batch automation can provide a high level of return if planned and implemented carefully. In certain situations it can form the basis for further integration of manufacturing systems or as a component of FMS. As has been demonstrated the approach is not limited to certain high technology sectors but can be applied across a range of industries.

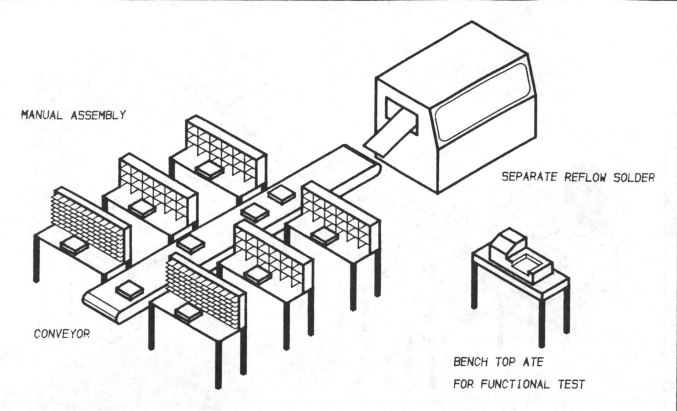

MANUAL ASSEMBLY

SEPARATE REFLOW SOLDER

CONVEYOR

BENCH TOP ATE

FOR FUNCTIONAL TEST

FIGURE 1

SPINE LAYOUT WITH PCB PALLETS FEEDING A FLOW SOLDERING MACHINE

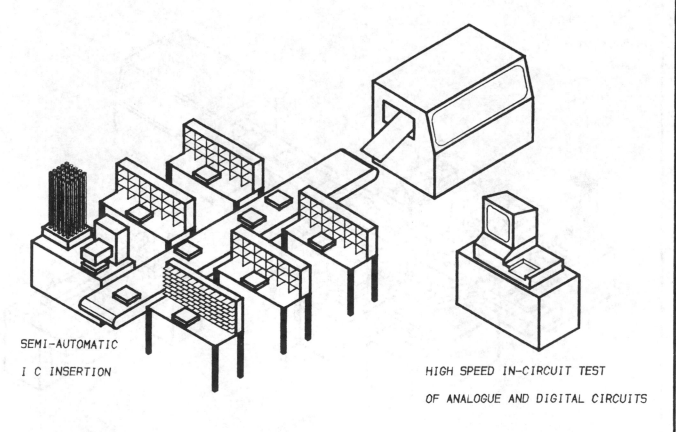

SEMI-AUTOMATIC

I C INSERTION

HIGH SPEED IN-CIRCUIT TEST

OF ANALOGUE AND DIGITAL CIRCUITS

FIGURE 2

USE OF SEMI-AUTOMATIC INSERTION AND TEST EQUIPMENT

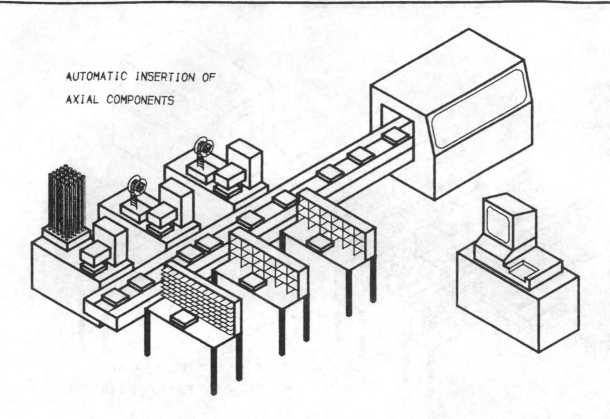

AUTOMATIC INSERTION OF
AXIAL COMPONENTS

FIGURE 3

FLOW LINE ASSEMBLY WITH INCREASING USE OF AUTOMATIC INSERTION

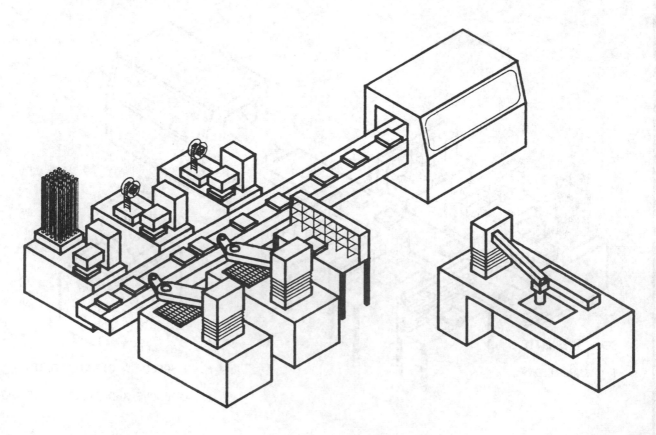

FIGURE 4

USE OF ROBOTS FOR ASSEMBLY AND TEST

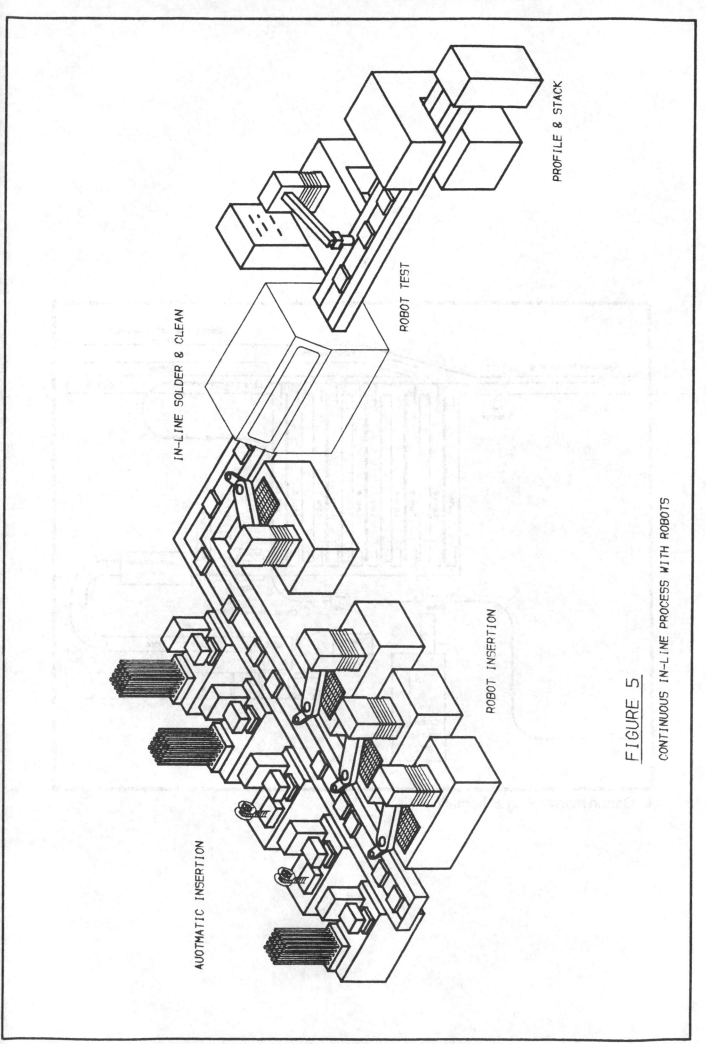

AUOTMATIC INSERTION

ROBOT INSERTION

IN-LINE SOLDER & CLEAN

ROBOT TEST

PROFILE & STACK

FIGURE 5

CONTINUOUS IN-LINE PROCESS WITH ROBOTS

127

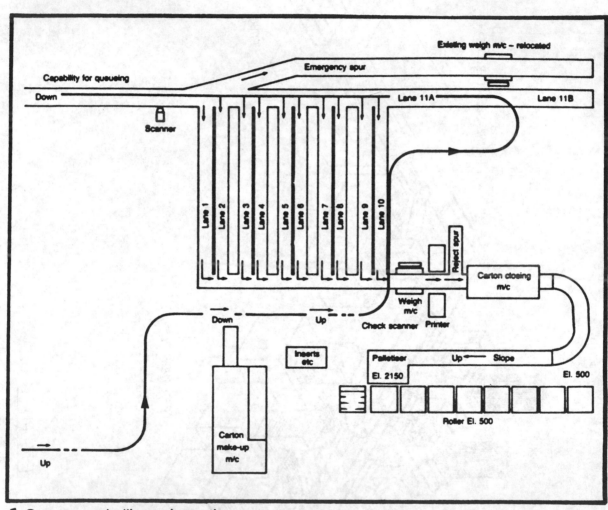

6 Carton marshalling schematic

PERSONNEL-SAVING, FLEXIBLE AND HIGHLY PRODUCTIVE DRILLING AND MILLING OF WORKPIECES FOR THE ELECTRONIC AND AVIATION INDUSTRY

Dieter Hofmann

Sales Director

HECKLER & KOCH

Maschinen- und Anlagenbau GmbH

D - 7230 Schramberg-Waldmössingen

1. INTRODUCTION

"All roads lead to Rome"

Both pragmatism and high intellectual flexibility can be recognized in this old quotation dating back more than 2000 years. Indeed, this kind of thinking can help even us modern humans out of many an intellectual cul-de-sac.

Pragmatism in production engineering primarily means the provision of what is necessary in economics terms when planning a plant or system and not what is technically feasible.

Flexibility in production engineering means adapting the production systems to the internal structure of the enterprise and to the external demands of the market.

The planning of modern industrial systems is not only affected by economics and engineering aspects but equally strongly by the qualifications of personnel, the availability of personnel and by the conditions dictated by the social environment.

It is quite clear that the market demands even now, and will demand
more so in the future, the possibility of supplying ever more
variants in ever shorter time spaces and all this to higher levels
of quality at simultaneously stagnating or even dropping prices.

It is precisely when investigating production costs that, as ever,
many of the "classical" flexible production concepts fail.

Turning such concepts into reality makes many and varied demands
on the organization considering the investment and these include:

a) high capital investments
b) high level of staff training
c) high level of organizational sophistication throughout
    the enterprise.

If one of the above conditions is not fully met, the economy
of the system immediately becomes doubtful. As far as small
and medium-size companies are concerned, the costs involved
reach levels which may in fact endanger their existence.

However, Messrs. HECKLER & KOCH Maschinen- und Anlagenbau GmbH
of Schramberg-Waldmössingen have in recent years proved on several
occasions that pragmatic thinking and the consistent evolution
of products can lead to considerable savings and other advantages.

This contribution introduces 2 machines which are characterized
by certain new concepts and which provide the user with important
potential savings.

The first machine is the HECKLER & KOCH BA 20 CNC DRILLING AND
MILLING CENTER which has found ready acceptance in many branches
of industry due to the high level of economy offered.

The second machine is a very new development, first introduced
during the German Machine Tool Exhibition in Stuttgart - 18th to
22nd September 1984 - where it was received with considerable
acclaim.

2. HECKLER & KOCH DRILLING AND MILLING CENTER BA 20 CNC

On approximately 60 to 70 % of all workpieces, the most frequent

130

machining operation is drilling, together with its related
operations countersinking, reaming and tapping.

Machining operations of this type have so far been completed in
a number of ways.

When producing small and medium batches on:

- pillar-type drilling machines
- in-line drilling machines
- machining centers

When producing large batches on:

- multi-spindle drilling machines
- special-purpose machines.

The above production methods have the following disadvantages:

- Manually operated pillar-type drilling machines are
  personnel-intensive (high cost)

- In-line drilling machines are fixture-intensive (long lead
  times, high costs, limited modification potential)

- Machining centers demand high hourly machine rates for
  relatively simple drilling operations

- Multi-spindle drilling machines involve expensive operating
  media

- Special-purpose machines offer no flexibility

Fig. 1   HECKLER & KOCH AUTOMATIC DRILLING MACHINE BA 20 CNC

The drilling and milling center BA 20 CNC developed by HECKLER & KOCH is
currently used by many companies for the machining operations described
because it advantageously avoids the disadvantages outlined. The range
of parts handled by these machines starts in small plants with batches
of 10 to 20 off followed by pre-production batches, employment in
departments of large plants and is finally concerned with mass production
where this automatic drilling machine is used in the automotive industry
for work which was hitherto the typical preserve of special-purpose
machines.

131

Fig. 2   BA 20 CNC machine configuration

The most important feature of the HECKLER & KOCH BA 20 automatic drilling machine is the possibility of operating it from two distinct positions. Operating the machine follows the motto that money can only be earned when a machine is producing chips. This is why the machine has two separate, rigid clamping tables. 12 tools are assigned to each of these tables and are automatically taken up by the work spindle under NC program control whereupon each set of tools can be applied to different workpieces. The drilling head proper is suspended from an extremely rigid gantry frame and reciprocates for the machining operations from one clamping table to the other. While fully-automatic machining of a workpiece is in progress on one side, the operator can unload, clean the fixture and reload the machine on the other side. After closing the guard door, the operator confirms completion of his actions. When the machine has completed the first workpiece, it traverses to the other machining area, automatically selects the machining program provided and now completes machining of the second workpiece, again fully-automatically. While these machining operations are carried out, the operator can once more remove the first workpiece and reload.

This machine working method combined with the versatile and flexible possibilities NC has to offer, has established a firm place for this machine in the production department of many plants and this in quite a short time.

Quite apart from the high level of productivity and output now possible with drilling operations, the HECKLER & KOCH automatic drilling machine is further characterized by the small amount of floor space it requires. Moreover, the machine build-up is exceptionally clear so that placing several machines in the care of a single operator presents no problems. A number of users have combined machines of this type to form production islands. A large user in Schweinfurt has made up a "cell" consisting of 4 machines looked after by a single operator. Another arrangement is a "mixed" cell consisting of a large machining center and an automatic drilling machine. This interesting configuration is employed for workpieces which undergo circular machining on the large machining center and are subsequently finished on the automatic drilling machine. We are here concerned largely with retaining holes which would have been far too expensive to produce on the large machining center.

Other companies employ the automatic drilling machine as a machine tool supplementing lathes. In such applications, the operator can look after both the turning machine and the automatic drilling machine which carries out subsequent operations. This is possible because both machines are automatic NC machine tools which only require the operator to carry out the loading and unloading functions.

The field of application for this automatic drilling machine is very universal on account of its favourable technical data. The two working areas offer in each case 650 mm along the X axis, 350 mm along the Y axis and 200 mm in the Z axis. The drilling quill is driven by a 10.5 kW DC motor which provides a maximum drilling capacity into St 60 steel of up to 20 mm diameter and is capable of tapping an M 16 thread - again into St 60 steel. The speed range covers 80 to 6100 r.p.m. so that widely varying materials can be machined. The rapid traverse rate along the X and Y axis is 12 m/min. so that positioning times are very short. The fast rapid traverse also assists with short tool changing times amounting to approx. 6 seconds.

When the economics of the BA 20 automatic drilling machine are compared with those of a conventional in-line drilling machine offering 20 mm diameter drilling capacity, the advantages of the new machine will become even more obvious.

The BA 20 automatic drilling machine is a machine tool with 24 spindles (tools). Compared with 4 in-line drilling machines with 6 spindles each, the automatic drilling machine would look like this:

a) Expenditure for personnel   Ratio 4:1
b) Floor-space required   Ratio 6:1
c) Investment involved   Ratio 4:1

Further economy features such as

- greater flexibility
- lower workpiece preparation cost and effort
- virtually no expenditure for special operating media
  improve the economy calculation even more in favour of the automatic drilling machine.

There are proven machining applications indicating advantages of far over 100 %.

The following machining examples again show quite clearly both the field of application and the potential improvements in the machining sequence and machining technology by employing the NC machine.

Fig. 3 - 7   Typical machining examples produced on
               BA 20 CNC.
               Comparison with conventional production
               methods.

The BA 20 drilling and milling center satisfies many expectations:

- high output because loading and unloading times are saved

- low personnel costs since several machines can be in the hands of one operator

- simple operation and programming because the machine configuration is extremely clear

- high and consistent workpiece quality because the machine works under CNC

- complete machining on a single machine is possible (drilling and milling)

- high level of reliability through use of few assemblies (for example: no tool changing mechanism because the tools are changed by the pick-up method)

- low floor-space requirement through very compact construction

The best possible proof of the many advantages the machine has to offer is likely to be the fact that more than 300 such machines have been sold in under three years.

Another example of how machines with completely new application potential can be developed from new ideas combined with the manufacturing experience gathered from using more than 100 CNC machine tools in in-house production, is the four-spindle horizontal machining center HBZ 16/4 by HECKLER & KOCH.

Fig. 8   4-spindle horizontal machining center HBZ 16/4

This machine is considered by many engineers as one of the most
remarkable machine tool developments of recent years and is
characterized by the following features:

- high productivity through simultaneous multi-spindle machining

- high production flexibility through up to 5-axis NC machining

- high workpiece flexibility through modular construction adaptable
  to particular problems

- high operational reliability through low number of assemblies

- high machine loading efficiency through classical NC operating and
  organization features

- high economy through
    • low personnel costs
    • low floor-space requirement
    • low energy requirement
    • no lost time for loading and unloading
    • short resetting times when changing workpieces

The machine is a horizontal planer-type machining center. High
output is assured by the provision of four parallel work spindles.

Fig. 9   4 parallel work spindles

The centre distance between the work spindles is 200 mm (optionally
250 mm) and this together with the total drive rating of 25 kW DC
will give some idea of the range of parts and the machining methods
applicable:

Fig. 10   Typical range of parts for HBZ 16/4

- Drilling, milling, countersinking, threading, reaming, boring
  (multi-spindle heads can be used)

- Machining of medium to large batches (close to output levels
  just under those considered as "classical" numbers for
  special-purpose machines)

135

It could be said that the typical application range is concerned with those batch sizes and workpieces which could be produced in a flexible and variable way on single-spindle universal machining centers, but not with adequate productivity. Another aspect would be that the total number of workpieces would be inadequate or not sufficiently certain to warrant the building of a special-purpose machine.

In the present-day situation, we are thus concerned not only with workpieces from the automotive industry but also with workpieces produced by accessory manufacturers. Parts from the electrical industry are just as suitable as parts from the aero-industry, precision engineering, optical industry etc.

Although the four-spindle version results in four-times the machine output (4 workpieces are machined simultaneously per work cycle), only 3 NC feed axes are involved.

The strict division of functions in the fundamental machine concept into working and moving section on the one hand and workpiece clamping section on the other (planer-type version or floor-type boring mill concept) permits modular construction and thus convenient adaptation to various workpiece ranges (shape, machining operations, clamping etc.). In this way, the machine can readily be incorporated as a unit into widely varying production systems and can be combined with many workpiece clamping and transfer arrangements (workpiece pallets, pallet pool, flexible transfer line, rotary transfer etc.).

Fig. 11   Separation of working section from
               workpiece carrier section

Tools with steep taper register ST 40 (automatic taper cleaning by compressed air is standard equipment, internal coolant supply is available as an option) are simultaneously exchanged in all four spindles with the aid of a linear tool changing arm.

The changing time for all 4 tools is approx. 10 - 12 seconds from cut to cut.

A chain magazine is installed separately and to the right of the machine, is readily accessible to the operator and can accommodate as standard a total of 100 tools. 25 tools are thus available for each work spindle.

Fig. 12   Loading and unloading while machining is
               in progress

In its standard version (X = 250 mm, Y = 680 mm, Z = 500 mm) the
machine works in conjunction with a remarkable workpiece carrier system.
The horizontal workpiece arrangement provides not only for exemplary
chip disposal but also allows simultaneous machining of 4 workpieces
secured adjacently to the horizontal swivelling workpiece carrier
beam. These workpieces can be simultaneously machined in a single
clamping set-up from at least 3 sides (or even 5 sides if a 4th
axis is added). However, the operator (or an automatic loader) can
also load and unload workpieces during the machining time because
one end of the workpiece carrier beam is always accessible for
loading and unloading. This workpiece carrier concept offers a
high level of flexibility (NC machining) combined with exceptional
universality (multi-sided machining) and high productivity (loading
and unloading create no idle times).

This new combination of successfully established features ensures
high savings, even in the standard version described, through high
output levels combined with high machine loading. The modular
construction of the machine permits convenient adaptation to
various workpiece ranges and production problems and/or to different
production concepts. Thus, for example, extending the X axis can
create a long-bed machining center which would then be suitable for
large workpieces in a 2-spindle or single-spindle version providing
it is equipped with one of a choice of workpiece clamping systems
(fixed clamping table, pallet pool etc.).

This, in other words, 3-axis machining unit can also be incorporated
in flexible transfer lines.

All these different versions provide a high level of reliability
and therefore a high degree of machine loading. This is the case
because despite the fact that four workpieces are completed
simultaneously, only 3 NC axes are involved (5 NC axes in the
most sophisticated version) in the machining operations together
with a CNC system (instead of 12 - 20 axes when employing
conventional single-spindle machining).

The same comments apply to operation, maintenance and organizational
incorporation of the machine.

Experienced operators, setters and service personnel capable of
looking after single-spindle machining centers are equally capable of
operating this machine without difficulty. No profound changes
to plant progress organization is necessary as would be the case
when introducing a typical flexible manufacturing system (FMS).

It is precisely these personnel and organizational consequences
which are frequently underestimated in the planning and operation
of typical flexible manufacturing systems (FMS).

Technical data - HECKLER & KOCH HBZ 16/4

Working traverses	X = 250 (1200) mm
	Y = 680 mm
	Z = 500 mm
Clamping area	4 x (300 mm x 1000 mm)
Drive rating	4 x 6.25 kW (total 25 kW)
Spindle speeds	60 - 4100 (80 - 5600) r.p.m.
Feed rates	1 - 4000 mm/min.
Rapid traverse rate	X, Y, Z = 15 m/min.
Tool magazine	Cassette chain
	4 x 25 tools (total 100 tools)
Tool changing time	Chip to chip for 4 tools
	approx. 10 seconds

3. SUMMARY

Rationalization and automation are essential to survive in worldwide
competition.

Modern production engineering offers many solutions. Technically
fascinating and complex major systems are a possibility for few
organizations only when looked at under stringent economics criteria.
However, pragmatically designed machine tools such as introduced
in this paper offer great savings to many organizations as they take
into consideration existing personnel qualifications and simultaneously
allow retaining the existing organizational structure.
Waldmössingen, 31st July 1984
HECKLER & KOCH
Maschinen- und Anlagenbau GmbH
Dieter Hofmann

Fig. 1

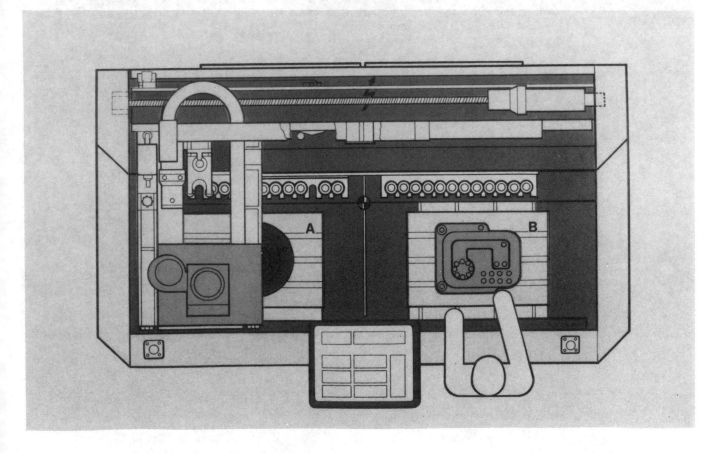

Fig. 2

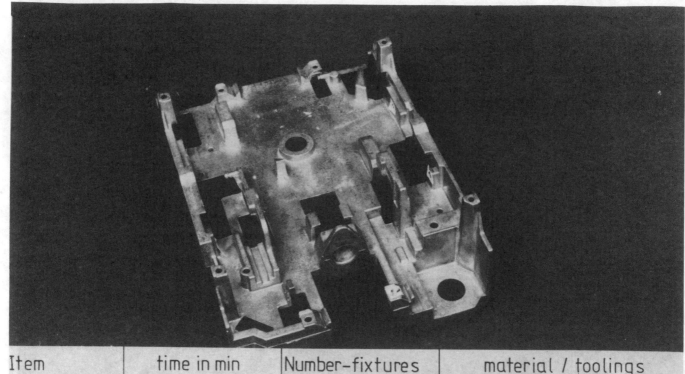

Item	time in min		Number-fixtures		material / toolings	
	conventional	Heckler & Koch	conventional	Heckler & Koch	conventional	Heckler & Koch
component (computer)	23 min	6,2 min	11	1	· milling m/c · gang drilling m/c · 3 milling attachments · 1 drilling attachment	· BA 20/CNC 781-2 · 1 fixture (multiple clamp- ing fixture)

Fig. 3

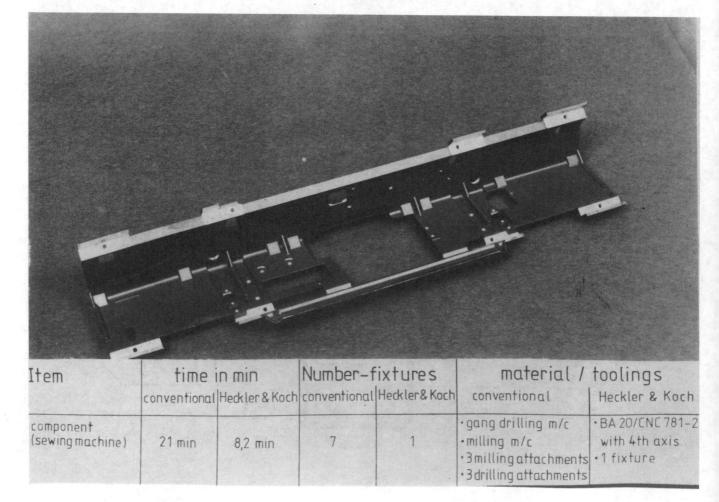

Item	time in min		Number-fixtures		material / toolings	
	conventional	Heckler & Koch	conventional	Heckler & Koch	conventional	Heckler & Koch
component (sewing machine)	21 min	8,2 min	7	1	· gang drilling m/c · milling m/c · 3 milling attachments · 3 drilling attachments	· BA 20/CNC 781-2 with 4th axis · 1 fixture

Fig. 4

140

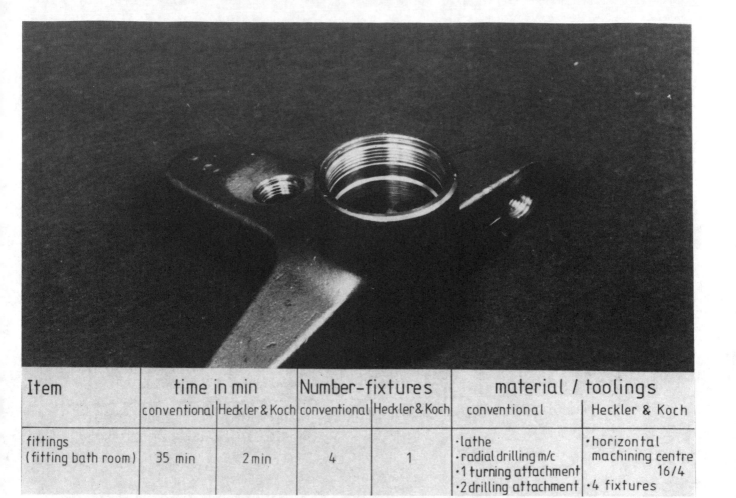

Item	time in min		Number-fixtures		material / toolings	
	conventional	Heckler & Koch	conventional	Heckler & Koch	conventional	Heckler & Koch
fittings (fitting bath room)	35 min	2 min	4	1	·lathe ·radial drilling m/c ·1 turning attachment ·2 drilling attachment	·horizontal machining centre 16/4 ·4 fixtures

Fig. 5

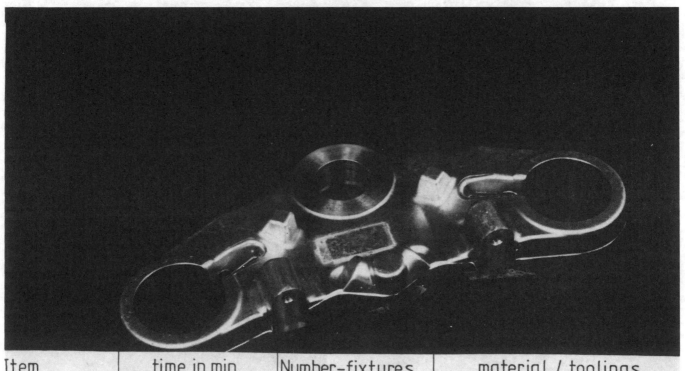

Item	time in min		Number-fixtures		material / toolings	
	conventional	Heckler & Koch	conventional	Heckler & Koch	conventional	Heckler & Koch
component (car)	7 min	3,5 min	3	1	·milling m/c ·gang drilling m/c ·1 milling attachment ·2 drilling attachment	·BA 20/CNC 781-2 ·1 fixture (multiple clamp-ing fixture)

Fig. 6

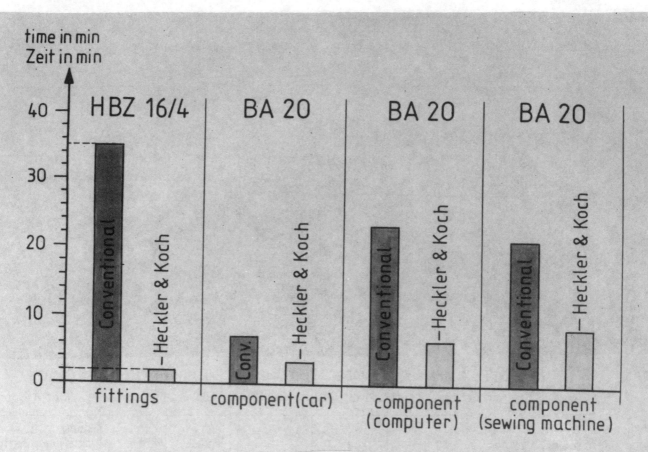

Fig. 7

Fig. 8

Fig. 9

Fig. 10

143

Fig. 11

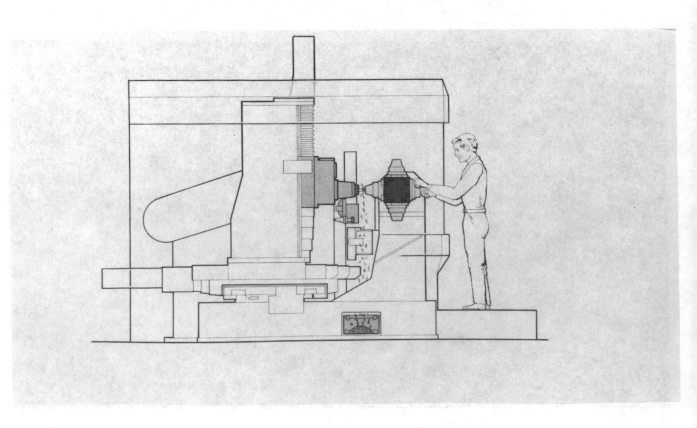

Fig. 12

144

TOOLING AND OTHER ASPECTS OF ADVANCED MANUFACTURING TECHNOLOGY

WHAT ROLE DOES THE CUTTING TOOL PLAY IN THE FULLY AUTOMATED MACHINING PROCESS

Address by Hans Sandstrom, Regional Product Manager, Sandvik
Trading Singapore Pte. Ltd., Singapore

Introduction

Today's modern CNC machines have been developed to give even
better stability and performance while the material and the
working allowance of the workpiece have been held within tighter
limits. Together these factors contribute to better utilization
of the cutting edge and shorter economic tool life. The end
result is more frequent tool changing. Let us go further
and see how the operation of the machine and the operation work
have changed.

Earlier the operation work could be divided into the following:
1. Operate the machine
2. Change workpiece
3. Change tools

Today the CNC is running the machine, robots or portal loaders
are changing the workpieces.
This means that an important part of the total down time is caused
by the cutting tools.
Another step further towards rationalization is automatic changing
of the cutting tools.
With this background in mind the Block Tools have been developed.

Modular Tooling System

When the Block Tool project was started in 1976, the objective was to create a modular system for automatic and semi-automatic tool changing, incorporating features such as high stability and repeatability, a simple clamping mechanism and high clamping forces. In other words, the coupling between the cutting unit and the clamping unit was the deciding factor.

In principle, two-thirds fo the conventional shank tool-holders become unnecessary because the unique Block Tool coupling provides the ability to achieve stable tool clamping in the machine. The front third, forming the small and lightweight cutting unit, can be regarded as the actual cutting tool, while the clamping unit remains permanently mounted in the machine.

The BT-System has the following features:

1. Stability

   As good as or better than a standard holder. The question of stability has had no. 1 priority for Sandvik for the following reasons:
   - A modular tooling system must not mean reduced cutting data because this can mean that a big part of the savings from automation is eaten up
   - Ceramics demands a high stability and ceramic is an important cutting material
   - The customers must not be limited regarding future development of cutting materials and geometries

2. Small Dimensions
   - The cutting heads are small and light enough to satisfy the demands of fast acting automatic tool changers and this feature is a great help also where heads are being changed manually

3. Repeatability accuracy
   - The repeatability accuracy of the BTS-coupling is needed where pre-measuring or presetting of cutting edges for finishing operations is to be done off the machine. The repeatability of the coupling is ± 5 microns in Z-direction and ± 2 microns in X-direction.

4. Clamping units - Size and operation

   The design of the coupling makes it possible to produce small clamping units with a very small drawbar stroke which means that it's very simple to operate manually.

5. Coolant supply

   The system must provide possibilities for enough coolant
   supply for short hole drills, etc

6. Complete assortment

   A complete assortment of tools in different sizes are
   available and special tools can easily be manufactured
   from blanks

7. ISO-inserts

   The system is designed in such a way that ISO inserts
   can be used. Similar as solid shank tooling.

   The features I have mentioned corresponds very well
   with the specification we planned back in 1976 when we
   started the development of the BTS.

   We have tested a number of couplings and at an early
   stage we had to decide whether we wanted to go for a
   cylindrical or square shank solution.

   We chose the square solution for the following reasons:

   - More rigid and smaller dimensions
   - Optimal clamping - pulling force in relation to the
     inserts
   - External square shank tools don't need radial clearances
     but makes it possible to use larger and thicker inserts
     (ceramics)
   - A tangential support can be used which gives the following
     advantages:
   - less overhang i.e. better stability
   - adjustment of centre height
   - the support can be used as a coolant distributor.

The result is a modular tooling system adapted for maximal
metal removal rate for external machining in combination
with sufficient space for chips.

Manual CNC lathes

For lathes without workpiece changing equipment, manually
operated Block Tooling is the first step towards rationalization. Such an investment incorporating a hydraulic
system for the clamping of cutting units would minimize the
tool changing and set-up time.

To prove to you that manual tool changing can be profitable

I want to show you one example from Alfa Laval in Sweden who has equipped two Okuma LC40 with BT with manual changing. The machines are used for production of stainless steel components.

The machines are used to two shift production. Manufacturing batches are small with 4-5 re-rigs per day. Previously rather heavy cassettes containing shaft-holders were changed on every re-rig. The objective with the BTS-installation was to reduce the re-rigging time giving shorter overall times and less capital tied up as a result.

Some facts

- 4-5 re-rigs/day
- 4-5 tools changed per re-rig
- 5 min. changing time per conventional tool
- 0,8 min. changing time per BT-unit
- Machine costs = 600 Swedish kronor/hour = 75US$/hour
  = 10 Swedish kronor/hour = 1.25US$/hour

Saving in time

- 5 X (5-0,8) = 21 min. per rigging
- 4 X 21   = 84 min. per day

Saving in money

- 84 X 10 = 840 Swedish kronor/day = 105US$/day
- 200 X 840 = 168.000 Swedish kronor/year = 21.000US$/year

Investment

- Clamping unit : 48.000 Swedish kronor/machine=6000US$/machine
- Cutting unit : 14.400 Swedish kronor/machine=1800US$/machine
  TOTAL   62.400 Swedish kronor/machine=7800US$/machine

Pay-off: $\frac{62.400}{168.000}$ ≈ 5 months $\frac{7800}{21000}$ ≈ 5 months

Fully Automated Lathes

The real step towards more effective machining is, however, the completely automated lathe. Full automation of tool changing is essential if a bigger number of tool geometries are requested and if the machine is running unsupervised for

any length of time. In general the economic returns available
from better utilisation during manned shifts and extra unmanned
running more than justify the extra machine cost for automation.
The more automatic the machine operation, the more important
the operating economics become. The investment cost increases
of course with automation though extra hours running more than
compensates this.

The costs for tool changing and measuring cuts however decreases
sharply. The highest possible cutting data should be used
consistant with predictable tool life. This element of tool
life predictability takes on a new importance in an automated
machine and argues strongly for systematic tool management and
specialised tool maintenance.

The best storage capacity if obtained with a cylindrical drum
magazine. The standard magazine has a capacity for up to
240 tools, depending upon the tool size. The tools are stored
in a magazine rack with room for 5 to 10 tools in each rack.
Because the racks are exchangable, with a simple manipulation,
handling of worn or new tools is very rational.

The Block Tool system also contains standardized grippers
with positive clamping of the tool to give maximum security.

The measurement probe is the next important link in the chain.
To check the positioning of the workpiece in the machine, a
measurement probe with a Block Tool coupling is used. The
Block Tool probe can be stored in the magazine during machining, so that it is protected from the hostile environment in
the machining area.

For the gauging operation the probe is coupled to a clamping
unit equipped with contacts which convey the measurement
signal to the control system.

To obtain a completely automatic lathe, monitoring of the
process is necessary. The Block Tool System includes a
Tool Monitor which is a micro-computer based unit for
monitoring signals from the feed motors, spindle motors or
sensors for measuring the feed force.

The Tool Monitor is programmed with the help of learning
control. After a learning cycle has been carried out, the

limits are set which for each respective operation will determine the condition of the tool.

We shall now go further and see how a considerably larger investment in a <u>fully automatic lathe</u> can be economically justified.

In this case we show how a customer has assessed an offer from a machine tool manufacturer.

The offer was based on a 4-axis CNC-lathe equipped with:
- a portal loader for workpiece changing
- automatic tool changing
- measurement probes for tool gauging
- a Tool Monitor for process monitoring.

Some facts

- Two shift production
- About 40 tool changes/day
- 3.5 min. changing time per tool inclusive of measurement cuts with conventional tools.
- 0.6 min. changing time for automatic changing including the probe cycle
- Cycle time/component = 0.9 min.
- 72.000 components/month to be produced.

Time saving in tool changing

40 X (3.5 - 0.6) = 116 min./day

Time saving due to automatic operation

(Lunch periods, etc)
2 X 60 = 120 min/day
Total time saving = 236 min/day

Available time in 2 shifts

Conventional CNC	575 min.
Automatic CNC	811 min.

Required machine capacity

Conventional CNC	6 machines
Automatic CNC	4 machines

150

Savings in Investments

6 Conventional CNC	9 MSEK	=	1.1 Mill. US$
4 Automatic CNC	8.4 MSEK	=	1.05 "
Saving	0.6 MSEK	=	0.05 "

Cost Saving

No. of operators per 2 shifts reduced from 6 to 2.

Operator cost saving	4 X 150.000SEK =	4 X 18.750 US$
Other variable cost saving	100.000SEK =	12.500 US$
Savings in variable costs	700.000SEK =	87.500 US$/year

In the highly industrialized world where we are living today the "AUTOMATION" is not just a buzz-word for the enthusiasts, its reflects a demand felt by almost production engineer related to machining for increased production and a high level of utilization of capital employed.

The cutting tool has proved to play an important role in metal cutting and the Block Tool System can be considered as a key-link in the fully Automated Machining Process.

BTS has been on the market for a couple of years now and
- more than 600 installations are running today in production with very satisfactory results
- approximately 50 fully automated machines are in order or have been delivered.

152

APPLICATION OF LASER BEAM CUTTING IN THE MANUFACTURING OF CUTTING TOOLS

* T.Nakagawa, * H.Yokoi, ** C.S.Sharma, * T.Suzuki, * K.Suzuki

* Institute of Industrial Science, University of Tokyo, Roppongi, Minato-ku, Tokyo 106, Japan
** Department of Production Engineering, Victoria Jubilee Technical Institute, Matunga, Bombay 400019, India

Laser beam cutting technique has been used for the manufacturing of shaving, broaching and milling tools. The good features of laser cut sheet laminated blanking die have been used to produce a multi-edged sheet laminated blanking-cum-shaving tool. The principle of multi-edged shaving is further extended to obtain a broaching tool to be used on a press machine. A successful attempt at cutting high speed steel sheet by laser beam has led to the production of a fly milling cutter for machining a profile. The experimental results of these cutting tools show great potential in the application of laser beam cutting to the manufacturing of such cutting tools in the near future.

1.INTRODUCTION

The idea of producing blanking tool by laminating thin hard sheets has recently been realized by cutting the contour of the die by laser beam [1]. This has been made possible by an accurate laser beam cutting machine equipped with CNC. Considering the high speed of cutting, the expected good accuracy of cut contour and the quality of cut, further work was carried out on the manufacturing of various kinds of blanking tools such as compound, drop-through type and progressive blanking tools[2,3]. The results of the blanking tests conducted on these tools are highly encouraging. Besides achieving a considerably high dimensional accuracy on the blanked product by such laser cut sheet laminated blanking tool, the cost of manufacturing is remarkably low and such a tool can be made ready within a short period.

Since the laser cut edge of the sheet is sufficiently sharp and hardened. It may be used for other metal cutting operations like shaving, broaching, milling and turning. Shaving and broaching tools appear to be more interesting for the

153

application of laser cutting, while milling and turning tools for profile essentially require laser cutting as a new approach of cutting tool production due to its fast cutting ability and low cost even for one piece of a tool.

In the trial of shaving and broaching tool, laser cut bainite steel sheets have been used as cutting edge. However the trial cutting of profile on high speed steel sheet proved to be successful by imparting smooth cut surface and sharp edge. Hence a laser cut profile to a fly milling cutter has been used for trial. Possibility of producing various turning tools has been also discussed. The present work particularly describes the design and production of shaving and broaching tools in detail.

2. CHARACTERISTICS OF LASER CUT EDGES

The capability of laser cut steel sheets to be used as cutting edges can be judged by examining the cross-section of a laser cut sheet surface. The cross-section views of laser cut surfaces of bainite and high speed steel sheet are shown in Fig.1 along with their surface roughnesses in the peripheral direction. The corner on the upper side, which corresponds to the entry side of the laser beam, is sufficiently sharp to induce cutting of sheet. This side of the laser cut surface has been used as the cutting edge in blanking and piercing, while the edge of the other side is spoiled due to dross and insufficient sharpness. This can easily be confirmed from the increased surface roughness on the periphery near the bottom corner of the cross-section in Fig.1.

The cut surface of high speed steel sheet also shows remarkably good cutting edge. It possesses sharp edge and comparatively low surface roughness, which may encourage the use of high speed steel sheet for making various kinds of laser cut sheet laminated. The laser beam cutting conditions for cutting bainite and high speed steel sheets are shown in Table 1. Because of the inavailability of high speed steel sheets with large size, bainite steel sheets have been used for both shaving and broaching tool, while a fly cutter of high speed steel cut is used for milling operation test.

3. LASER CUT SHEET LAMINATED SHAVING TOOL

3.1 Basic Principle

Shaving is a secondary operation which is carried out mainly for the purpose of finishing the blanked contour surface. The mechanism of shaving is more or less near to the cutting action generally observed in machining. Therefore a small amount of material — usually left as shaving allowance uniformly all over the contour during blanking — is to be removed in the shaving operation. However, very large amount of material must be removed to get a required cut surface in the case of rather thick sheets. In such cases, generally two or more number of shaving steps are needed. A combined shaving with multi-cutting edge can be realized by using the idea of sheet lamination, which is otherwise very difficult to achieve in usual method of tool making. The basic principle involved in the construction of the proposed shaving tool is illustrated in Fig.2.

In this figure the evolution of a shaving tool is illustrated by steps (a) through (d). The first step indicates the lamination of a number of laser cut sheets in which the edge of the first (top) sheet represents the blanking die edge and the subsequent edges are for shaving operation. The idea is to perform blanking followed by subsequent shaving in the same down stroke of the press ram. Therefore the difference between the two consecutive edges is the amount of material removed during shaving per edge, i.e. the shaving depth or depth of cut (λe). The total amount of material to be removed, i.e. is the difference between the blanking edge and the last shaving edge. The punch is also proposed to be made by laminating the slug portion left in laser cutting of sheets for die making. Since the punch has to pass through the last shaving die edge also, a clearance 'Cls' is to be kept between them. This clearance, however, is possible to achieve by cutting the contour twice during laser cutting. As a result of laminating such sheets with lser cut contours, the tool may be named as multi-edged blanking-cum-shaving tool.

154

The manufacturing of tool by this method seems to provide greater flexibility in the design and manufacture of tool to achieve certain specific characteristics. Such flexibility can be explained by the illustrations from (b) through (d). The first requirement of a multi-edged cutting tool is to make provision for the space for chip accummulation and the regular disposal of chips. The chip pockets can be easily made by cutting similar contours — but larger than the neighbouring cutting edge — on steel sheets which can be used as spacers as indicated in Fig.2 (b). The laser cutting of such spacers can be performed by an offset program installed in the NC laser machine. Considering that the chip disposal is to be performed by compressed air, it is also possible to make necessary provision for keeping hole passage in the laminated structure to blow off the accummulated chips after each cycle by pressurized air. The scheme is illustrated in Fig.2 (c).

Other important features of a laminated blanking-cum-shaving tool may be the possibility of changing the blanking clearance and depth of cut per shaving edge. This is illustrated in Fig.2 (d). According to the sheet thickness to be processed, the blanking clearance can be changed by reducing the number of laminates. On the other hand the depth of cut per shaving can be reduced by replacing the previous laminates with more number of thin laminates and dividing the total depth of shaving among them. The same is true in the opposite way to increase the depth of cut. Such changes may be needed depending on the variation in material and thickness of the sheet to be blanked and to be subsequently shaved.

3.2 Construction of the Tool

The prototype blanking-cum-shaving tool has got an inverted type blanking and shaving die arrangement as illustrated in Fig.3. The punch and the stripper are mounted on the lower shoe and the multi-edged laminated die is mounted to the upper shoe along with an ejector pad operating in the return stroke by the urethan bushes compressed during the down stroke. The punch, the ejector and the stripper are also made by laminating laser cut sheets because of convenience and cost saving.

The component to be blanked and shaved is a tensile test piece. This component shape has been chosen with a view point of providing an easier method of manufacturing such test pieces which generally need lengthy and costly operation. The edge of tensile test piece requires not only fracture free surface but also work-hardening free surface. The important features which may draw the attention of tool designs are as follows:
(a) A standard depth of shaving cut is 0.04 mm and the maximum number of sheet lamination is eight.
(b) The depth of cutting can be set to 0.08 mm or 0.12 mm by choosing suitable combination of sheets.
(c) For blanking die cutting edge 2.0 mm thick bainite steel sheet and for shaving die cutting edge 1.6 mm thick bainite steel sheets have been used.
(d) The increased hardness of the surface in the heat affected zone (HAZ) is very much suitable for shaving operation.
(e) The ejector pad has been used to avoid distortion of workpiece due to large depth of shaving per edge and blanking clearance.
The appearance of the blanking-cum-shaving tool with its upper and lower shoe assembly separated is shown in Fig.4.

3.3 Experimental Results

The shaving experiment was conducted on the above mentioned prototype tool by blanking and shaving 2.0 mm thick aluminum sheet. Judging from the surface characteristics of the shaved component, it can be said that just laser cut surface consists of considerably rough surface zone also. This surface characteristics are easily imprinted on the component surface being shaved. In an attempt not only to improve the surface roughness during shaving but also to give sharp cutting edge, the rough portion of the finishing edge surface was polished by an oil stone (or diamond file) in which a layer of approximately 0.01 mm is removed. By using such finished shaving edge, a remarkably good surface finish on the shaved surface could be achieved as shown in Fig.5.

Since one of the important purposes of shaving is also to remove work-hardened zone from the blanked component, a test related to this fact has been conducted. The test is the usual tensile test in which the test was discontinued before complete fracture. The specimen with the area of necking enlarged and shown in Fig.6 indicates the origination of normal ductile crack in the centre rather than on the side edges. This confirms the absence of work-hardened zone on the shaved edge due to its complete removal in shaving operation.

4. LASER CUT SHEET LAMINATED BROACHING TOOL

4.1 General Outline and Principle of Proposed tooling
Broaching tool uses the method of successive removal of material by a series of cutting teeth arranged along its length. The total amount of material to be removed for getting a specific geometry on the component is divided into suitable steps, and the area of stock corresponding to each step is removed by one tooth decided at the time of broach design. The two consecutive teeth of a group have a difference in their level or profile sizes across the axis, which is generally termed as rise per tooth. The following considerations are necessary in the design of a broaching tool.
(i)   Gradual change in the height of tooth
(ii)   Gradual change in the tooth profile across axis
(iii) Staggering in pitch of the teeth
(iv)   Provision of chip breaker grooves
The above considerations required in the design and manufacture of broaching tools result in difficult manufacturing procedure, and hence each broach needs separate design and manufacturing procedure except few standard broaches. Since each tooth and pitch are different, machining operation becomes expensive and time taking. A geometrical complex broach, which is to be manufactured in small number, is bound to be very expensive.
A broaching tool can be manufactured by laminating laser cut sheets of specified tool material. The idea has emerged from the experimental work done on laser cut sheet laminated shaving tool. The principle involved is almost same as that of shaving. The principle of proposed broaching tool is illustrated in Fig.7.

4.2 Design and Construction of the Prototype Broaching Tool
The blank shape is shown at Fig.8 (a) and, by broaching both inner and outer surface, the final shape as shown at Fig.8 (b) is to be achieved. The total depth of cut in broaching this profile is 5 mm. The scheme of material removal corresponding to different sets of teeth are given in Table 2. The spacers of varying thickness are used to make changes in the pitch. The spacers to provide needed pitch are of different thickness such as 2.6 mm, 2.0 mm, 1.2 mm and 0.8 mm respectively for first, second third and fourth set of teeth. No spacer is used in between burnishing teeth in the final stage. The total height of the laminated cutting edge sheets and spacers is to be 173 mm. The proposed structure of the tool is shown in Fig.9.

4.3 Layout for Laser Cutting
In the broaching tool each tooth profile will have to be different, which should directly determine the modes of material removal to get a final product profile. In order to obtain cutting profiles of each sheet, we utilized cutter compensation function installed in NC machine. By applying this compensation function to the two fundamental geometrical data (Fig.10—A and B), we have gotten the two removal patterns of materials as illustrated in Fig.10—(a) and (b) for gear tooth profile.

4.4 Broaching Trial
The broaching experiment for 10 mm thick aluminium ring blank was conducted by using the assembled broaching tool set on the hydraulic press. The photographs of the assembled broaching tool and component after broaching are shown in Fig.11. Though fracture free surface could be obtained, several stick

slip marks were left on the cut surface. Further the cut surface of the finished component showed not so good surface roughness as Rmax around 10 μm. The roughness of the component corresponds to that of laser cut surface at exit side. If the users want to have more fine surface roughness, the rough portion of the laser cut surface should be removed by subsequent polishing operation.

In the manufacturing of the broaching tool by this method, several technical problems can be pointed out; namely, i) inclination of the axis of the assembled broaching tool from the normal direction, ii) weakness to the lateral load because of the sheet-laminated structure, iii) difficulty to realize an ideal chip pocket space between the laminated cutting edges, and so on.

5. MILLING AND TURNING TOOLS

Both edge and surface qualities of the laser cut high speed steel sheet (Fig.1) were satisfactory, and hence a fly milling cutter could be realized. To eliminate subsequent grinding for the relief face, the position of a slit for the laser cut sheet was shifted upward as shown in Fig.12. This fly cutter could be successfully used to cut the profile on an aluminum block (Fig.13). In case of giving positive rake angle to the cutter, the groove has to be shifted downward, though the subsequent edge grinding is necessary. The sharpness of the profile machined reflects the high speed steel as potential cutting tool for various machining operations. Particularly even when a small number and only one or two tools of specific are required, the laser cut tools are economically viable and are quickly manufactured.

A further application of this method in making cutting tools can be observed in the manufacture of flat and circular tools. In the case of flat form tool, it may be possible to provide relief angle by cutting the profile in tilted position.

6. CONCLUSION

On the basis of the successful performance of the blanking tool made by laser cut bainite steel sheet edge, the idea has been extended to the application of such cut edges to manufacturing of cutting tools. The trials are conducted for three types of cutting tools and the results are confirmed to be encouraging. The concluding remarks concerning this work may be arranged in three groups as follows:

(I) Cutting Edge
(a) The edge produced by laser cutting lies in the area of HAZ which is supposed to be a drawback of laser cutting. But such edges of the bainite steel sheet showed excellent toughness along with an increased hardness. Besides bainite steel sheet, the laser cutting trial made on high speed steel sheet also resulted in a very good cut edge which could successfully be used for machining.
(b) The laser cut surfaces are rather rough so that a good surface finish by using this cutting edge can not be expected. However, it has been found that polishing the cut surface with an oil stone or a diamond file can easily improve the surface finish to a great extent.
(c) In the laser cutting, it is difficult to provide either rake angle or relief angle to the cut edge. In spite of this, the cutting or shaving could be performed successfully without much difficulty.

(II) Application
(a) As an extension of laminated blanking tool, a single stroke blanking-cum-shaving tool with multi-cutting edges obtained by laminating laser cut bainite steel sheets has been realized. Since the shaving is performed by a number of cutting edges arranged in series providing small depth of cut, the shaved surface finish is superior to the conventional press shaving.
(b) Further, the principle used to shaving is extended to the design and manufacturing of a broaching tool which requires complicated and time

taking manufacturing procedure. This could again be realized by laser cutting of bainite steel sheets. A push type broaching tool, using a cutting action similar to shaving, has been designed and manufactured to be used on a press.

(c) As a result of the successful manufacturing of above cutting tools by laser cutting, the method is further used to manufacture a fly cutter for milling operation. Such a tool has performed machining of a profile successfully during trial.

(III) General Characteristics

The other good aspects which may be pointed out in favour of laser cutting are cost and speed. The manufacturing speed is high because both the main contour and the holes for fastening, dowels etc. can be made at one setting within a short period as compared to other manufacturing methods. This leads to drastic reduction in the manufacturing cost of cutting tool.

To cope with the design procedure involved in cutting tool like broaching, fly cutter etc., the CAD/CAM technique may be applied successfully with the laser cutting. Further more, this method enables economical production of small batch of cutting tools.

ACKNOWLEDGEMENTS

The authors wish to thank deeply to Suruga Seiki Co.Ltd. for their kind help in conducting the laser beam cutting, and also to Nisshin Steel Co. Ltd. for their kind cooperation in providing bainite steel sheets for die making.

REFERENCES

1. T.Nakagawa, K.Suzuki and K.Sakaue,"Manufacturing of a Blanking Tool by Laser Machining", Proc. of the Int'l Laser Processing Conf. Anaheim, (1981,11)
2. H.Yokoi, T.Suzuki, K.Suzuki, and T.Nakagawa, " Manufacturing of Blanking Tool and its Die-set by Laminating Laser-Cut Steel Sheets", Proc. of NAMRC -XII, (1984,5)
3. H.Yokoi, T.Suzuki, K.Suzuki, and T.Nakagawa, "Laser Cut Blanking Tool with Sheet Laminated Structure ", Proc. of the 5 th ICPE, pp.484-489, Tokyo (1984,7)

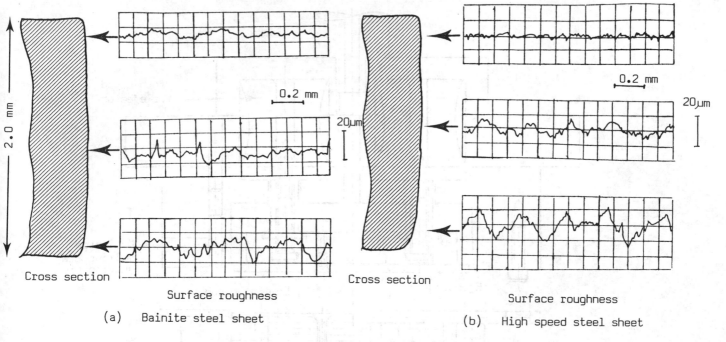

(a) Bainite steel sheet

(b) High speed steel sheet

Fig. 1   Cross-section views and surface roughnesses of laser cut
surfaces of 2.0 mm thick bainite steel sheet and high
speed steel sheet

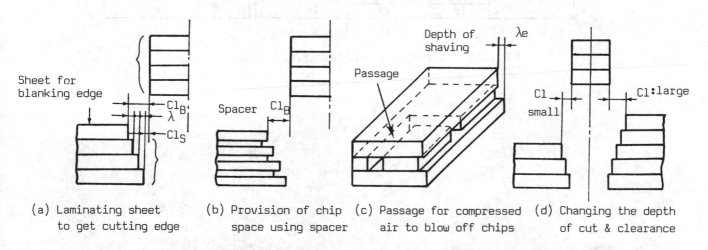

(a) Laminating sheet
to get cutting edge

(b) Provision of chip
space using spacer

(c) Passage for compressed
air to blow off chips

(d) Changing the depth
of cut & clearance

Fig. 2   Evolution of a laser cut sheet laminated shaving tool

Table 1 CO_2 laser cutting conditions for 2.0 mm thick bainite
steel sheets and high speed steel sheets

	Average power (Watts)	Cutting speed (mm/min)	Pulse width (msec)	Frequency (Hz)	Assisting Gas
Bainite steel sheet (t2.0)	250～300	500	1.0	300	O_2 (4.0kg/cm^2)
High speed steel sheet (t2.0)	250～300	500	1.0	400	O_2 (4.0kg/cm^2)

Laser cutting machine: Miyama-Urawa's cutting device mounting
Coherent's Everlase M46

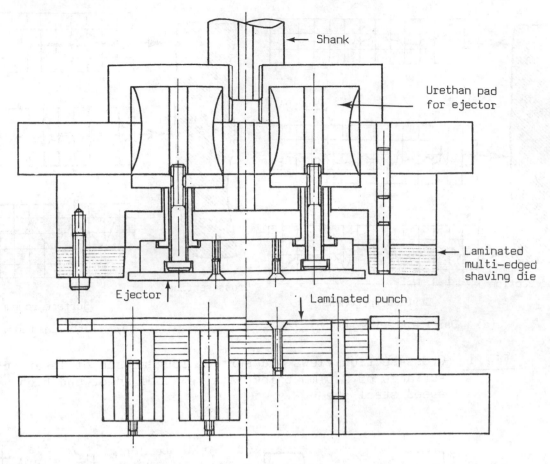

Fig. 3   Structure of laminated shaving tool

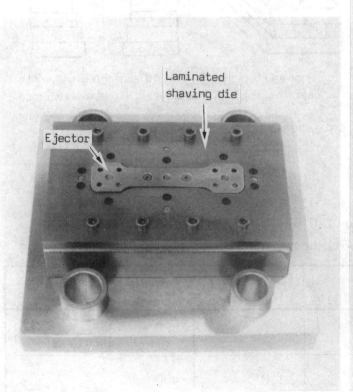

(a) Upper part of the shaving tool

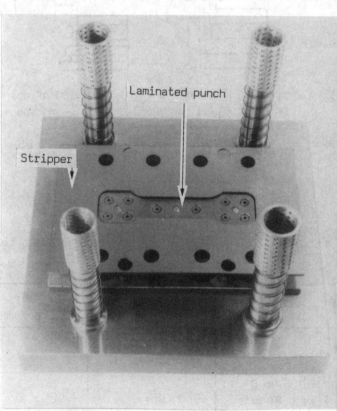

(b) Lower part of the shaving tool

Fig. 4   Appearances of the prototype shaving tool

160

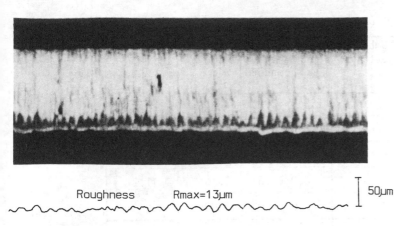

Roughness   Rmax=13μm   ‖ 50μm

0   1   2 (mm)

(a)   Shaved surface directly by laser cut edge

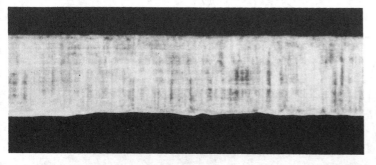

Roughness   ‖ 50μm

0   1   2 (mm)

(b)   Shaved surface after polishing the laser cut surface of sheet

<u>Fig.5</u>   Comparison of the surface conditions shaved by laser cut edge directly and by laser cut edge polished with oil stone

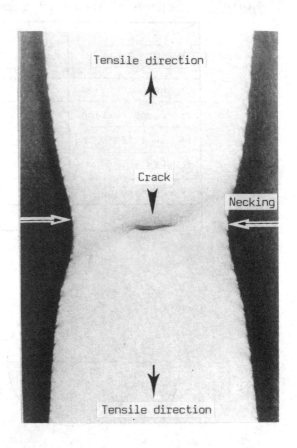

<u>Fig.6</u>   Normal ductile crack appeared in the center of the blanked and shaved specimen by this tool, which shows the removal of work-hardened zone (2.0mm mild steel sheet)

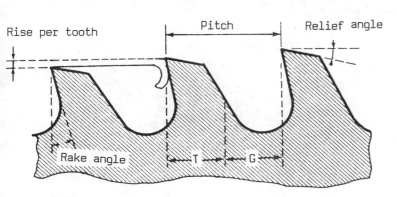

T : Tooth thickness and back up portion
G : Chip space

(a) Tooth shape of conventional broach

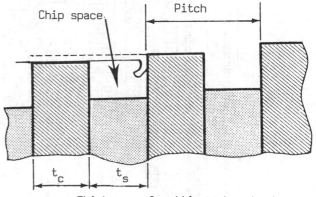

t_c : Thickness of cutting edge sheet
t_s : Thickness of spacer sheet

(b) Tooth shape of laminated broaching tool

<u>Fig. 7</u>   Principle of a proposed broaching tool by laminating laser cut sheets

Table 2   Scheme of material removal (sheet material: bainite steel sheet)

		Rise per tooth (mm)	Number of teeth	Thickness of teeth (of spacer) (mm)	Stock removal (mm)	Height (mm)
I	Rough cutting	0.15	20	2.0 (2.6)	3.00	92
II	Rough cutting	0.12	12	2.0 (2.0)	1.44	48
III	Semi-finishing	0.08	5	2.0 (1.2)	0.40	16
IV	Finishing	0.04	4	2.0 (0.8)	0.16	11.2
V	Burnishing	0.00	3	2.0 (0.0)	0.00	6
	Total	–	44	–	5.00	173.2

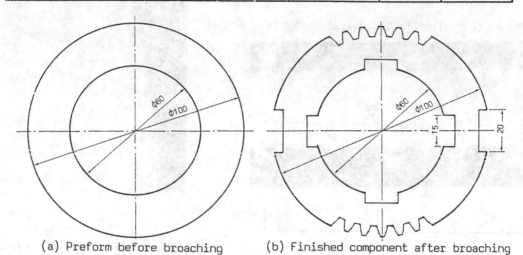

(a) Preform before broaching   (b) Finished component after broaching

Fig.8   Component for the prototype broaching tool

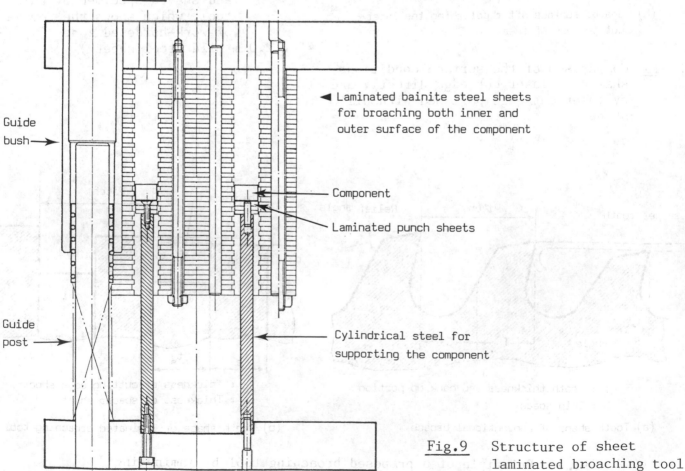

Guide bush→

Guide post→

◄ Laminated bainite steel sheets for broaching both inner and outer surface of the component

Component

Laminated punch sheets

Cylindrical steel for supporting the component

Fig.9   Structure of sheet laminated broaching tool

162

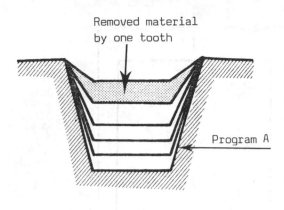

Removed material by one tooth

Program A

(a) Rough cutting stage for gear tooth profile

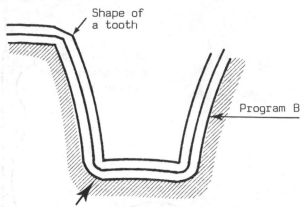

Shape of a tooth

Program B

Final shape of the component

(b) Finishing stage for gear tooth profile

<u>Fig.10</u>   Two fundamental patterns of material removal

Laminated cutting teeth

Guide bush

(a) Upper part of the prototype broaching tool

Punch sheets

'b) Lower part of the prototype broaching tool

(c) The component after broaching
(10 mm thick aluminum)

<u>Fig.11</u>   Appearances of the experimental sheet-laminated broaching tool and the component after broaching

163

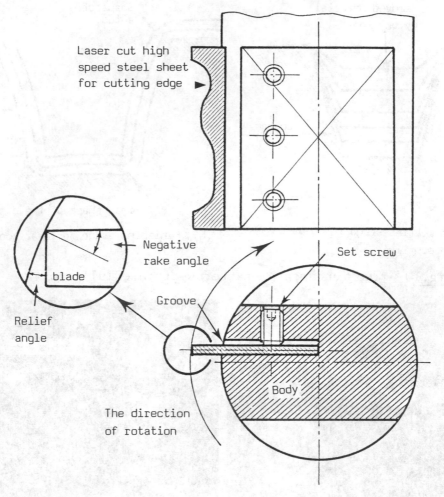

Laser cut high speed steel sheet for cutting edge ►

Set screw

Negative rake angle

blade

Relief angle

Groove

Body

The direction of rotation

Fig.12   Structure of the fly milling cutter constructed by laser cut high speed steel sheet

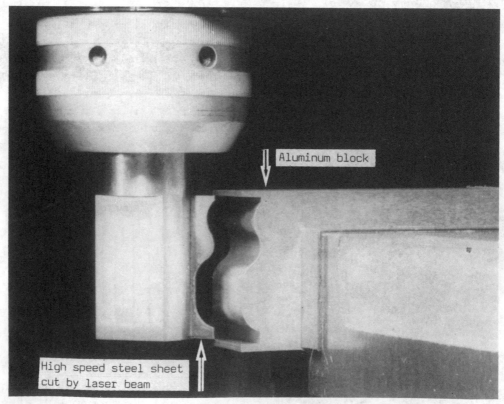

Aluminum block

High speed steel sheet cut by laser beam

Fig.13   Appearance of the machined work piece and the fly cutter set up to a milling machine

CHIEF SALES EXECUTIVE

AXEL FRIEDMANN, GERMANY F.R.

THE MODULAR FIXTURING SYSTEM, A PROFITABLE INVESTMENT!

Delivery-deadlines and the demand for rationalization in fixture making. - the   (1)
fixture room, the scapegoat of production.

Common production judgements:

non-productive, expensive, slow, non flexible, - a milestone around the neck.

It is, however, worthwhile to consider production reality: new production orders
are most likely accepted based on already exceeded delivery dates with the fixture
room being the last link in the production chain, which is supposed to compensate for
any time lost. Or, how often does it happen that conventional fixtures have to be replaced immediately following break-down in production. To combat all these date pressures and also to endeavour rationalization, modular fixturing systems represent a real
alternative.

Modular fixturing systems

Picture 1

165

are fully accepted along with conventional fixtures, i.e., they can be considered
technically equivalent, yet they feature striking economical advantages, as this
presentation will outline. These advantages occur in various areas such as reduction
of fixture stock and thus capital engagement, the reduction of machine numbers and
last not least the reduction of labour cost in fixture design and manufacturing.
What is the reason for the ever increasing application of fixturing systems in the
past ten years? They have been on the market for more than twenty years without being
recognized. Only with the entry of NC-techniques into our plants changes in attitudes
took place. This happened at a point when NC machines were suddenly no longer prestige
objects and their production costs influenced calculations directly. In recent years
the development of such manufacturing systems has been pushed forward in a comet like
manner. New developments reduced manufacturing times, settingup times were decreased
progressively. Improvements were literally "squeezed" from the systems with the result
that each manufacturer of machining centres was delighted when his tool changer reduced
chip to chip time by second or so, compared with his competitor. Only after the machines had been "squeezed" to their limits the periphery was considered and now modular
fixturing systems were remembered. Today these are an essential part of efficient
production. Two systems with different connecting principles have been established.

Connecting principle (2)

a) Slot system

Picture 2   Picture 3

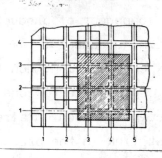

This system appears to be most widely spread in manufacturing and application. The
basic idea follows the example of slotted machine tool tables, combined with support -
and clamping elements, as workholding systems. The main characteristics of the slot
system are the connection of individual modules with matching T-units. The assembly
takes place on a slotted base plate which is manufactured very accurately and extremely wear resistant due to complete hardening. The force transmission between slot and
T-unit takes place in different ways for each co-ordinate axis:

    X-axis: frictional locking plus form locking

    Y-axis: form locking

    Z-axis: frictional locking

These basic considerations are of great importance to the fixture designer, who would
normally be inclined to transmit the forces acting on the fixture from the component
by means of form locking. The advantage of this system is the option to position

modules freely in one direction thus permitting a favourable adaptation to the component geometry.

b) Hole system

Picture 4

Picture 5

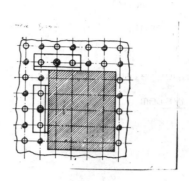

Characteristic for the hole system is the prevailing application of bolts and dowel pins - or just bolts for lower accuracy requirements - as connection elements. The initial idea was to avoid the costly manufacturing of slotted elements by using circular matching surfaces, which are generally produced less expensively. With this system it is essential that the matching holes are absolutely co-axial and their pitch dimensions are within close tolerances. For this, two manufacturing alternatives are available:

The hardened bushes may either be pressed into the hole which has been produced on a jig-bore, or, match bushes are cast into pro machined holes using a master mould. The disadvantage is that only the bushes are hardened. The base plate, which is to the same degree as the bushes subject to wear and also has significant influence on the component accuracy, is unhardened. Due to the alternating arrangement of precision and tap holes a large matrix dimension is created, making the hole system relatively inflexible. It is hardly possible to achieve form locking when locating the parts to be machined.

Elements - Slot - system.   (3)

Picture 6

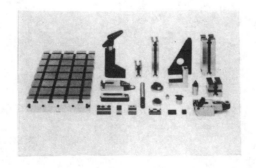

Following, the most commonly used elements of modular fixturing systems are introduced, which may basically be divided into the following groups:

1.) Base plate
2.) Mounting blocks
3.) Connection elements
4.) Locating elements
5.) Clamping elements
6.) Tool guidance elements (drill bush holder)

<u>Elements - hole system</u>   (4)

Picture 7

The elements of this system are divided into the same function groups as for the slot system.

<u>The coupling technique of HALDER-Slot systems.</u>   (5)

Picture 8

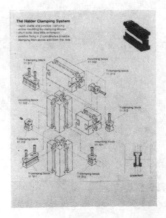

168

Picture 9

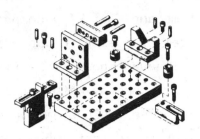

Introduction of a fixturing system in production.   (7)

In most cases the initial status appears as follows:

   1.) production costs too high

   2.) production flow too inflexible, due to manufacturing times for conventional production equipment. Resulting objective for toolroom and fixture room:

   1.) significant reduction of fixture costs

   2.) more flexible fixture back-up

a) Initially the following questions have to be raised:

- Is the purchase of a fixturing system economically justified?

- When, or after how many fixturing systems does the pay-back start?

- Are the components in question long running series or are modifications to be expected?

- Is multi-side machining required in one set-up?

- Will the base plates, provided with a fixturing set, accommodate the components regarding size and shape?

- How complicated are the parts?

- Will the fixtures be used only once or repeatedly, and if, how often?

- How do the costs of dedicated fixtures, completing the same requirements, compare?

- What are the costs for assembling and dismantling the fixturing systems?

- What do repetitive costs amount to?

- What is the ratio of positioning and clamping times between fixturing systems and special purpose fixtures?

- Are the required elements available during the application term of the fixturing system?

- Will the demands on the system be satisfied with existing modules or would additional elements have to be manufactured?

- Could a fixturing system be utilized for different operations following slight modifications?

- Which fixturing system suits the requirements of the component range in question?

- How many fixtures are expected to be assembled from the fixturing system per year?

- What is the expected amortization period for the system?
- Can appropriately qualified personnel for the assembly of such systems be recruited?
Following the answering of these questions, including consideration of assembling and dismantling costs, application time and cost of a compatible special fixture, the decision whether a fixturing system or spezial purpose fixture is commercially justifiable can be made. The commercial advantages of the fixturing system are particularly obvious when numerous plants (e.g. a Company with various subsidiaries) utilize a centrally located system, in which case the system would be located in the place with the most frequent usage. This plant could rent the assembled fixtures for a fee (following experience this could amount to 12 - 15 % of the cost of a special fixture). The rent period averages 10 days. This procedure influences efficiency and rentability favourably.

<u>Costs of a fixture assembled from a modular fixturing system.</u> (8)

In order to establish a feasibility study, the total costs of the considered and alternative in the form of a fixturing system have to be indentified. This is essential for a feasiblity study and comparsion with a special purpose fixture. According to the following formula, established by Dr. Brueninghaus, the cost of a mudular fixture can be indentified as follow:

Picture 10

$$MF_{tot} = N(1+Y) C_{ass} + C_{MF_c} =$$

$$TU \left(\frac{1}{T_W} + \frac{IR}{2} + 0.05 \right) + C_{MF_s} \times T_N + C_{C_{man}} \quad (DM)$$

<u>Nomenclature:</u>

MF_{tot}	=	Total cost of modular fixtur	(DM)
N	=	Number of different required fixtures	(number)
Y	=	relative number of repetitive assemblies	(number)
C_{ass}	=	Assembly costs for one modular fixture	(DM)
C_{MF_c}	=	Cost contribution of modules	(DM)
T_U	=	Average utilization of fixture	(days)
T_W	=	Writing off period	(10 years)

C_{MF_s} = Storage cost for modular
system. (DM)

T_N = Period for cost
consideration (years)

$C_{C_{man}}$ = Production costs of
components (DM)

IR = calculated interest rate (DM)

The best way to establish feasibility is to compare a modular fixture with a purpose built fixture, granted the modular fixture is applicable.

Case study for a metal cutting operation producing gear boxes:

(purpose built gear boxes in batches)

Picture 11

Hourly rates:

Fixture design	DM 52,--/h.
Fixture room	DM 60,--/h.
Q.C.	DM 48,--/h.
Modular fixture system assembly	DM 54,60/h.

Number of assembled fixtures per year:

Total number of new modular assemblies	184
Number of repetitive assemblies	32

Cost of one fixture:

Average cost of a dedicated fixture DM   2.630.--

Costs for assembly and dismantling of a modular´fixture:

(3 h + 0,6 h x 54,60) DM   196.56

Repetitive assembly and dismantling costs for a modular fixture

1.8 h x 54.60 DM   98.38

Comparison of annual total costs:

Total cost of 184 special fixtures

184 x 2.630,-- DM 483.920,--

Total cost of 184 new and repetitive modular fixtures:

184 x 196.56

+ 32 x 98,38 DM 39.315.20

Annual savings DM 444.604,80
=============

Based on a total investment of DM 100.000,-- (modular kit for 5 - 8 medium sized
fixtures plus working place) an amortization period of only a few months will be
achieved with profits exceeding the investment several times.

<u>Organization</u> (9)

Picture 12

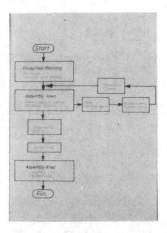

In order to secure success of such an investment a suitable organization, apart fom an
innovative attitude, is essential. Picture No. 12 shows simplified organization scheme,
which would obviously have to be adjusted to any company specific requirements. It is
very much advisable to keep the assembly area clear from personnel not involved as
well as easily cleanable.

Picture 13

The organization flow takes place as follows: The job planning department processes
order including component drawing and job sheet to fixture assembly area. It is advisable to provide a component at this stage. Location, - fixation, - and clamping
points are indicated in the drawing. Subsequently the fixture setter selects the
required elements and builds the fixture literally around the component without the
requirement of providing a drawing beforehand. Following the return from production
the setter establishes an assembly plan prior to dismantling

Picture 14

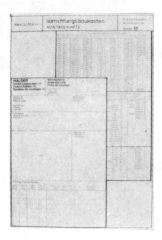

and also memorizes the assembly by taking an instand picture. This guarantees that possible alterations e.g. during production are memorized for future repetitive assemblies. The assembly plan is then recorded as a base for repetitive assemblies.

Final note:

Following many years of experience it can be established that the introduction of a modular fixturing system is no longer a question of rentability but a question of an innovative attitude.

Fixture examples:

Picture 15   Picture 16

Literature: H. Matuszewski

"Modular Fixturing Systems"

technica 23/1982

G. Brueninghaus

Computer Aided Design

of Modular Fixturing

Systems

174

Photochemical Machining - A Cost Effective Rival to Metal
Stamping for Parts Manufacture

Dr. David M. Allen, Lecturer and Consultant in Photomechanical
Engineering, Cranfield Robotics and Automation Group, School of Production
Studies, Cranfield Institute of Technology, Bedford MK43 OAL, England.
Also Technical Liaison Member to the Photo Chemical Machining Institute,
Pennsylvania, USA.

Photochemical Machining (PCM) is an advanced manufacturing
technique used mainly in the USA, Japan and Western Europe. Although
similar chemical etching techniques can be utilised to fabricate
integrated circuits from silicon, printed circuits from laminated foils
and optical gratings and graticules from coated glass, PCM as an industry
confines itself predominantly to the manufacture of parts from relatively
thin sheet metal. The PCM process, its technical and financial
considerations and its applications are described.

INTRODUCTION

Photochemical machining (PCM), also known as photoetching, is a multi-
stage manufacturing process employing photographic and photoresist
technologies with wet chemical etching being used to dissolve away
material from specified areas of the substrate.

PCM is admirably suited for producing complex piece parts in relatively
thin sheet metal. The metal products are found in the electronic,
automotive, aerospace, telecommunication, graphics, decorative, computer,
optics, medical, nuclear, metal-working and precision engineering
industries.

Although PCM was developed over 25 years ago, it is only in the past
decade that engineers have begun to recognise it as an extremely useful
manufacturing method. Photoetching techniques have now been refined to
the extent that some products, such as colour T.V. receiver tube aperture
masks (see Applications) can be made only by this method.

PROCESS DETAILS

The basic PCM process is outlined in Figure 1 and involves:

1. Production of the desired registered patterns on photographic films or glass plates, known as the phototool.

2. Coating of the cleaned sheet metal with a light sensitive film of photoresist. The photoresists used are usually negative-working (i.e. rendered insoluble in a developer after irradiation with actinic light) and are supplied either as a liquid, requiring the substrate to be dip-coated and then dried, or as a dry film which has to be laminated onto the sheet metal by a combination of heat and pressure.

3. Contact printing of the phototool image on to the photoresist using ultra-violet light as the actinic exposure source.

4. Formation of durable stencils on the metal by development of the photoresist.

5. Etching through the stencil apertures. Industrial etchants are based on aqueous ferric chloride or cupric chloride solutions with carefully controlled additions of hydrochloric acid. (1)

The edge profile of the etched hole changes with etching time (Figure 2). Once the etchant has dissolved the top surface of the metal, and penetrates below the level of the stencil, etching can occur equally in any direction. The lateral etching is responsible for the appreciable undercut (Figure 2(a)) produced. A quantitative assessment of the resultant etched profile is given by the term 'etch factor' which relates the undercut (U) to the depth of etch (D) according to the equation:

$$\text{etch factor} = {}^D\!/_U$$

Etching through mirror-image, registered stencils can produce convex, straight or concave profiles dependent on the etching time, as illustrated in Figures 2(b), 2(c) and 2(d). It can also be seen that straighter profiles are obtained with less undercut when etching from two sides rather than one side (Figure 2(e)). Etching through dissimilar, registered stencil apertures may be used to produce tapered profiles (Figure 2(f)). This feature of PCM is probably unique in metal cutting techniques.

Dimensions, as well as profiles, alter with etching time. Hole and slot dimensions increase and outside dimensions decrease as etching progresses. Therefore, in order to produce a dimensionally accurate component with the desired edge profile, it is necessary to determine the etching time which will give the required profile. The etched features are then measured and, if necessary, a new dimensionally compensated phototool is produced which can then be used for subsequent production runs. Details of this procedure have been published by the author elsewhere (2).

TECHNICAL CHARACTERISTICS

As a general rule dimensional tolerances in production runs can be kept down to ±10% of the metal thickness down to ±0.025mm. As centre-to-centre dimensions are independent of etching time the associated tolerances reflect the accuracy of the phototooling. Typically, the tolerance is ±0.05% of the centre-to-centre dimension of the finished part e.g. 50±0.025mm.

In comparison with fine blanking and piercing (the most precise form of stamping), photoetched parts are burr-free, thereby eliminating secondary finishing operations and their chemical and physical properties are identical to those of the original sheet metal from which they are made. This is important where stress, high temperatures or loss of magnetic permeability must be avoided during manufacture. It should also be remembered that the alternative process of laser cutting produces heat affected zones at all cut edges - an undesirable feature in the manufacture of some components.

When making profiles of the types illustrated in Figure 2, the deviation from a perfectly straight profile is usually controlled to better than 20% of the metal thickness unless, of course, a tapered profile is required.

Table 1 compares technical characteristics of PCM with those of stamping.

FINANCIAL CONSIDERATIONS

Because hard tooling does not need to be used in PCM, lead times are remarkably short due to the ease of phototool manufacture. Typically, lead times are days rather than the months associated with stamping. Short lead times give the entrepreneur an opportunity to market his product sooner and gain customer acceptance before rival products enter the market.

If both PCM and stamping can manufacture a product to the same technical specification then the most economic production method needs to be ascertained. PCM involves low cost tooling but high costs per part, whereas stamping produces parts at a low cost but tooling costs are very high. For any part a break-even quantity may be determined, below which PCM is the more cost-effective production method and above which stamping is more economic. The break-even quantity depends on part complexity for, as shown in Figure 3, as complexity increases the break-even quantity rises sharply.

APPLICATIONS

Colour television receiver tube aperture mask

Commonly known as a shadow mask, this perforated component is found between the arrays of red, blue and green phosphor stripes on the inside of the T.V. screen and the three electron beam sources which activate them.

Figure 4 shows typical dimensions of the tapered slots etched in 0.15mm thick mild steel. The mask comprises 300,000 slots each of which must be perfect. The profile is obtained as illustrated in Figure 2(f). As customer demand for the product is vast, the masks are made by a highly automated, continuous 24 hour process in order to eliminate batch to batch variation and to provide increased output of the necessary quality.

Integrated circuit lead frame

A lead frame is an example of an electronics component with a complex geometry. High volumes are required but as integrated circuit designs change so rapidly, PCM is favoured as a production method because lead times are so short, thus allowing the I.C. to come onto the market quickly.

Magnetic recording head laminations

Laminations made from magnetic materials such as HyMu 80 are etched, rather than stamped, because stamping adversely affects the magnetic permeability of such materials.

As well as manufacturing these products together with many others relating to the electronics and electrical industries such as heat sinks, hybrid circuit pack lids, and light-chopper discs,(3) PCM is used to fabricate parts for a wide range of engineering applications (e.g. shims, washers, diaphragms, filters) and is being utilised increasingly in the decorative and graphics industries to produce signs, labels, sculptures, models and jewellery. (4)

Surface etching allows logos, instructions, part numbers and other alphanumeric characters to be incorporated into designs while etched fold lines enable a third dimension to be added to flat components for the fabrication of boxes and enclosures (Figure 5).   It can be seen therefore that the product range is vast and that within the limitations set out in Table 1, it is said that if the designer can draw it, PCM can make it!

CONCLUSIONS

Consideration of the technical and financial aspects of PCM enables one to conclude that PCM is:

1. the best method for production of prototypes (modifications to parts being accomplished easily by rapid low cost changes in phototooling)

2. excellent for small and medium batch production

3. excellent for volume production of complex parts such as lead frames

4. the only method suitable for production of extremely complex parts such as colour television receiver tube aperture masks

5. economically viable for the production of both simple and complex components in difficult-to-stamp materials such as molybdenum

REFERENCES

1. Allen, D.M., Hegarty, A.J. and Horne, D.F., Surface textures of annealed AISI 304 stainless steel etched by aqueous ferric chloride - hydrochloric acid solutions; Trans. Inst. Metal Finishing, 59, 25-9, 1981.

2. Allen, D.M., Horne, D.F., Lee, H.G. and Stevens, G.W.W., Production of spring steel camera shutter blades by photoetching; Precision Engineering, 1, 25-8, 1979.

3. Allen, D.M., Photochemical Machining; Electronic Production, 11, No. 9, 113-7, 1982.

4. Allen, D.M. Photochemical Machining in the U.K. (A review of the period 1980 - 82). PCMI Journal No. 14, pp 4,5, Fall 1983.

	PCM	Stamping
Material thickness	1.5mm (6mm for low resolution work)	13mm
Deviation from a straight profile	<20% of material thickness (T)	A slight taper with a burr
Minimum aperture size	Ø = 1.1T for most metals (but not an absolute limit)	Ø = 0.5T (low carbon steel) Ø = 0.75T (high carbon steel)
Material	All metals (but vary in etchability)	Non-brittle metals
Process advantages	1. Produces burr-free and stress-free components 2. Physical and chemical characteristics of metal not altered during processing 3. Variable edge profile	1. Forming operations can be carried out whilst blanking 2. Fast
Process disadvantages	1. A multi-stage process 2. Thickness limitation	1. Long lead times 2. Deburring required

Table 1. Technical characteristics of PCM in comparison with stamping

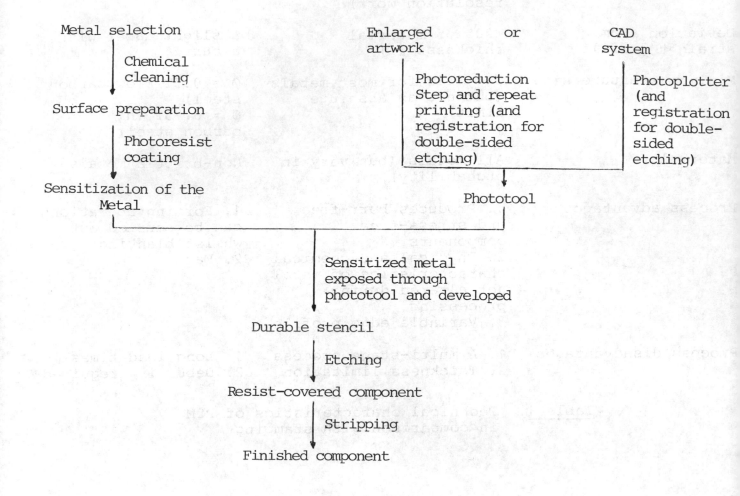

Figure 1.   PCM process stages

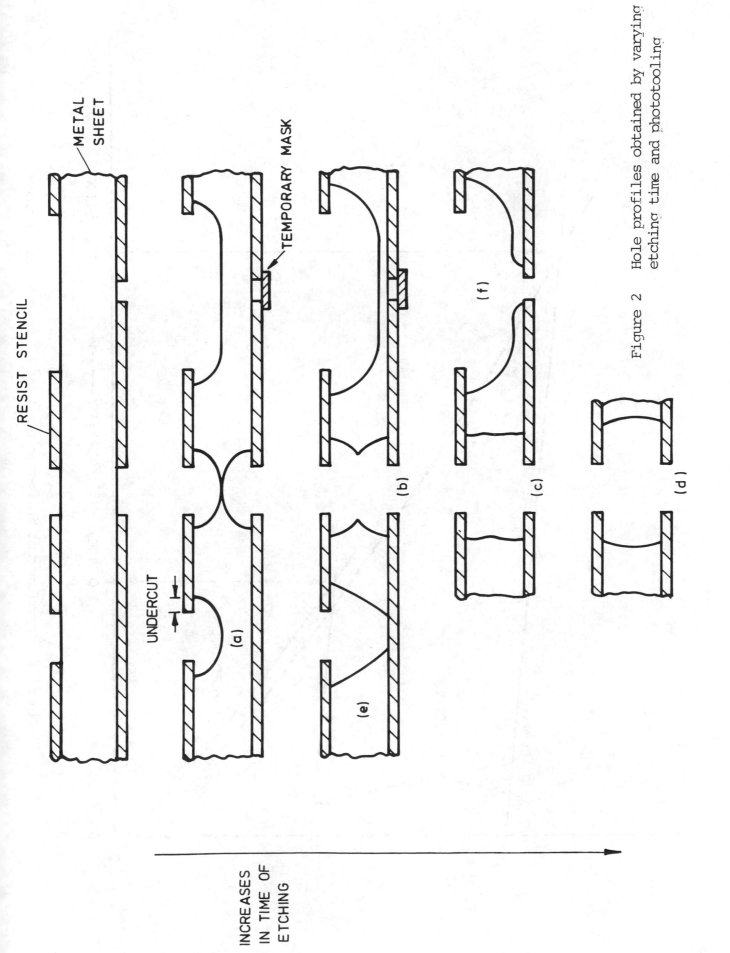

RESIST STENCIL

METAL SHEET

TEMPORARY MASK

UNDERCUT

(a)

(b)

(c)

(d)

(e)

(f)

INCREASES IN TIME OF ETCHING

Figure 2   Hole profiles obtained by varying etching time and phototooling

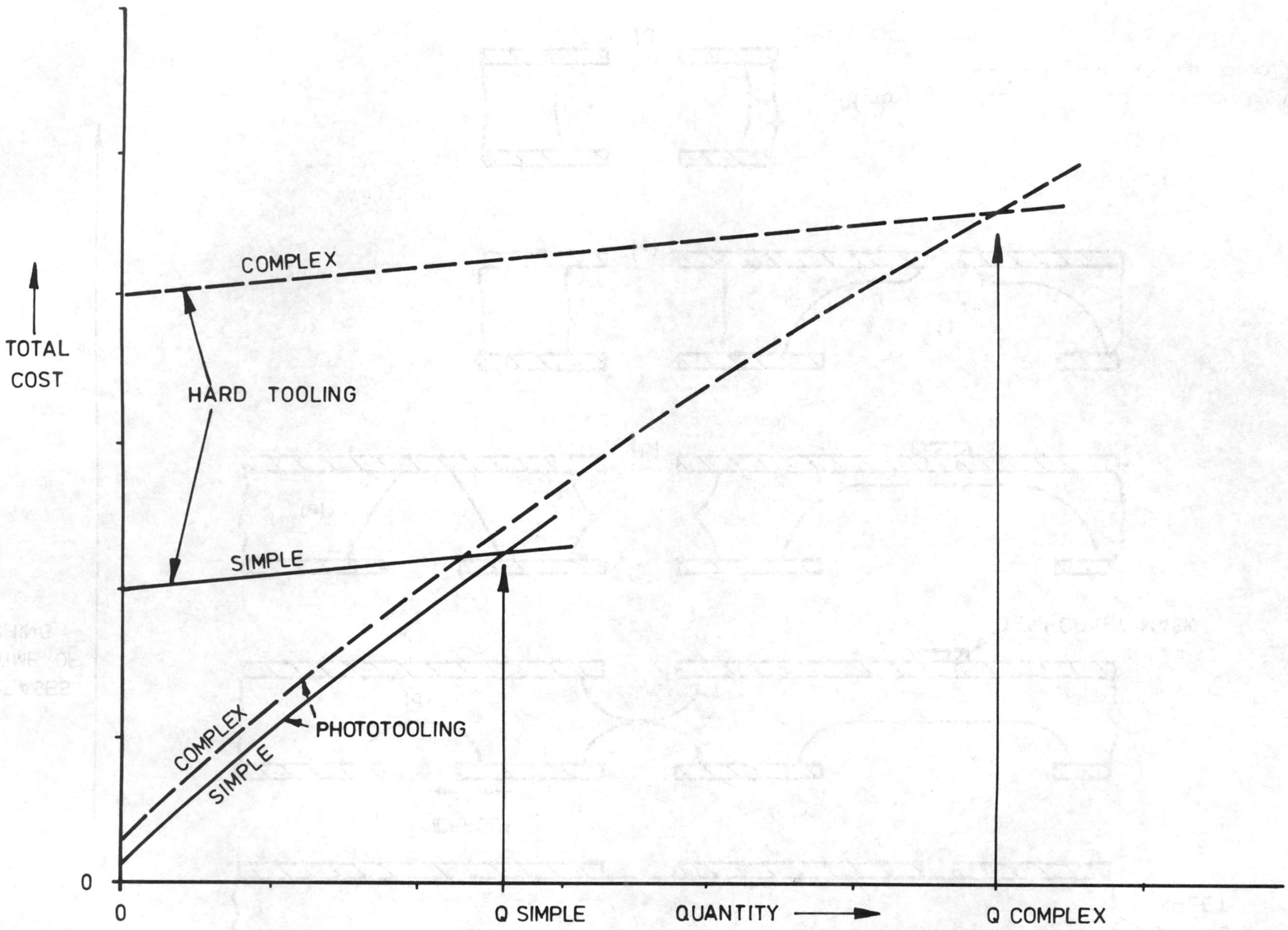

Figure 3   Dependence of break-even quantity (Q) on part complexity

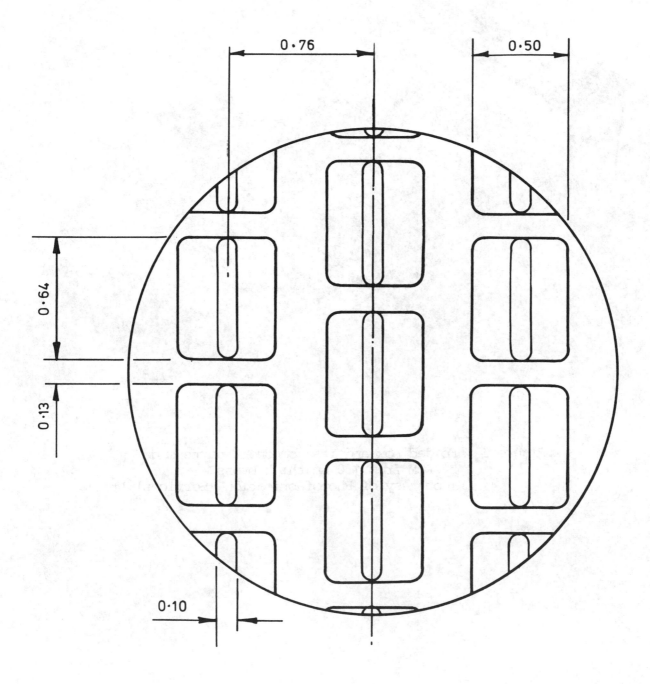

Figure 4   Typical dimensions (mm) of tapered holes etched through
0.15 mm thick mild steel to form a colour T.V. receiver
tube aperture mask.

Figure 5   Folded box and its component parts made
by PCM from 1.0 mm thick brass.
[Courtesy of Photofabrication (Services) Ltd.]

FOOTNOTE: Actual costs of part production based on figures supplied by Tecan
Components Ltd, UK.

185

TOLERANCE:
± 0.1 mm ON
0.5 mm METAL

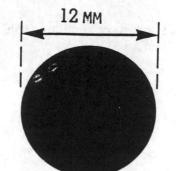

12 mm

	TOOLING(£)	PARTS(£)	TOTAL(£)
100 PARTS			
STAMPING	20	7	27
PCM	45	20	65
5000 PARTS			
STAMPING	100	85	185
PCM	65	90	155

TOLERANCE:
± 0.1 mm ON
0.5 mm METAL

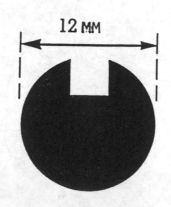

12 mm

	TOOLING(£)	PARTS(£)	TOTAL(£)
100 PARTS			
STAMPING	180	7	187
PCM	50	20	70
5000 PARTS			
STAMPING	300	85	385
PCM	70	90	160

TOLERANCE:
± 0.1 MM ON
  0.5 MM METAL

3 Stage
Progression

	TOOLING(£)	PARTS(£)	TOTAL(£)
100 PARTS			
STAMPING	450	10	460
PCM	85	20	105
5000 PARTS			
STAMPING	800	100	900
PCM	95	100	195

TOLERANCE:
± 0.1 MM ON
  0.5 MM METAL

Multi-stage
Carbide
Progression

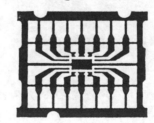

	TOOLING(£)	PARTS(£)	TOTAL(£)
100 PARTS			
STAMPING	25,000	30	25,030
PCM	300	55	355
5000 PARTS			
STAMPING	25,000	300	25,300
PCM	320	600	920

Flexible Automated Warehousing System

B.N. Karpov, V.A. Korolev, S.M. Sergeyev

Worldwide experience in using FMS has shown that one of the
most important links of these systems is the automated
warehousing system (AW). AW operation determines the main
parameters of FMS, such as production range of products,
which influence the flexibility degree of the whole enterprise and determines the main technical-and-economic indices.
In the paper discussed are the results of studying the
flexible rearranging AW with various degrees of rearrangement for the undetermined goods flow existing in FMS. Re-
commendations are given concerning the AW structures and
elements which, in the process of developing FMS, enable to
substantially increase FMS efficiency.

The wide employment of the flexible manufacturing systems in the world allows to generalize their component elements in order to shape their structure, in which the automatic warehouse (AW) is one of the most important elements greatly determining the indices of the production as a whole. According to the information furnished by the manufacturing companies, an average factory uses up to 40% of industrial areas for warehousing facilities at high cost, while their volumetric space is used far from fully. Besides, the actual time of treatment of the articles makes up some 5% and less, whence it follows that the tempo and character of the AW operation set limits on the productivity of FMS, output capacity and diversity of nomenclature, which influence the degree of flexibility of the whole production and the basic technical-and-economical characteristics.

While the problem of raising the flexibility of the basic equipment of FMS is solved as a whole by both the development of the software and great diversity of the treating tools and fixtures, the automatic warehouses at present are busy mainly with perfecting algorithms of operation of the robot-stacker which has a certain limit conditioned by the rigidly determined structure of the great majority of AW constructions.

So, in order to bring the degree of flexibility of automatic warehouses nearer to other elements of FMS, it is necessary to ensure the possibility of flexible rearrangement of its design to suit the parameters of goods to be stored. The research and analysis of contemporary warehouses prove that at present, taking into account the technical capabilities, the most rational solution is the realization of automatic warehouses with cells of rearrangeable dimensions. For efficient development of such systems, it is necessary to elaborate methods of calculation of their basic indices with due regard to main peculiarities of operation as a component of FMS.

Automatic warehouses designed and operated at present represent a system of multi-cell racks served by a robot-stacker, the dimensions of the cells being determined during designing and not changed after construction of the warehouse. The volume V of goods kept in the warehouse is variable and the limits of the range of its change are conditioned by the capacities of the equipment used in FMS (machine tools and the like) and auxiliary facilities (transport, robots and the like), and are known beforehand, which can be expressed by the relation $V_{min} < V < V_{max}$. Besides, the volumes of goods in the flow of supply to such warehouses are very stochastic values, which can be characterized by the density of distribution which is also influenced by such basic factors as the orientation of the production, the equipment used and also the production program, the control executed over the process of the production program selection being that particular lever you can use to influence the indices of FMS as a whole and the operation of the automatic warehouse in particular.

The described model of a rigid warehouse will serve as a starting point for the analysis. It is evident that in such a warehouse the volume of cells corresponds to the maximum warehousing of goods V_{max}, besides, due to elaboration of optimum control algorithms, several units of goods of smaller volume can be warehoused in one cell. Fig. 1a shows n - number of unitary goods in a cell, versus volume V (stepped curve D_1); also represented therein is another characteristic D_2 corresponding to the ideal variant of arrangement, featuring multiple volumes of goods in warehousing, while any characteristic of arrangement in a rigid warehouse will be always below D_2.

For estimation of indices of an automatic warehouse we use the data on distribution of volumes in the goods flow. Assuming, that $V_1 = V_{max}$, $V_2 \ldots V_p = V_{min}$ are abscisses of steps on the characteristic curve of arrangement D_1 (Fig. 1a), then C_N, the number of cells required for arrangement of N units of goods will be calculated from the formula:

$$C_N = \sum_{i=1}^{P-1} entier \left(\frac{N}{i} \int_{V_{i-1}}^{V_i} P_v(v) dV \right) \tag{1}$$

For curve D_2 of optimum arrangement of goods we have the values:

$$C_N = ent \left(\frac{N}{n^*} \int_{V_{min}}^{V_{max}/n^*} P_v(v) dV \right) + \sum_{i=1}^{n^*-1} ent \left(\frac{N}{i} \int_{\frac{V_{max}}{i+1}}^{\frac{V_{max}}{i}} P_v(v) dV \right) \tag{2}$$

It is evident that $C_N \geqslant \overline{C}_N$.

On the basis of the values obtained one can determine the most important parameter of the warehouse, that is the volume utilization factor, which is numerically equal to the value of relation of the busy volume of cells, where the goods are warehoused, to their total volume, and designated as K_v.

Taking into account (1), one has for the rigid warehousing (index "r").

$$V_{max} K_v^{(r)} = V_N / C_N V_{max} \tag{3}$$

where $\qquad V_N = N \int_{V_{min}}^{V_{max}} V P_v(v) dV \qquad$ is the natural value of N goods in the

warehouse, while in estimating the value $K_v^{(r)}$ it is necessary to use the relation:

$$K_v^{(r)} \leqslant \overline{K}_v^{(r)} \quad, \quad \text{where} \quad \overline{K}_v^{(r)} = V_N / \overline{C}_N V_{max}$$

It should be noted that K_v depends on $P_v(v)$, V_{min}, V_{max}, n^*, besides, it depends to a certain extent on N, however with great values of N its influence is so insignificant (for example, with $N \geqslant 100$, K_v changes less than by 2.5%), that it can be neglected in most cases. Further, with the normalized distribution commonly used, the influence is exerted only by the value of the relation

$$q = V_{max} / V_{min}$$

Passing over to the consideration of a flexibly rearrangeable warehouse, it is impossible to plot a dependence similar to that shown in Fig. 1a, because there is no constant volume of cells, however to enable the comparison of the indices of a flexible and a rigid warehouse, one can refer to a number of units of goods arranged in a unit of volume corresponding to a single cell of the rigid warehouse, though this number is not always an integer.
For quantitative description of discreteness and the degree of rearrangement and hence the character of the construction solution of the warehouse, we introduce the value m - index of discreteness of a flexible rearrangeable automatic warehouse which is numerically equal to the number of possible rearrangements depending on the inner size of a single cell of a rigid warehouse, while to simplify the calculation, we shall assume that the rearrangement is effected in equal steps.
In this case one can plot in Fig. 1b an analogous dependence for the relation between the volumes of goods and the number of units in a rigid warehouse. The stepped curve D_2 corresponds to concrete volume value m, the curve D_1 characterizes the limit case of an ideally

flexible AW (i.e. $m=\infty$) is described by the equation $n\,V = V_{max}$, which means that the volume of the busy cells is equal to the volume of goods.

Due to the fact, that in Fig. 1a the stepped curve D_1 corresponds to one realization of the probabilistic flow of incoming goods, by averaging such dependences in accordance with the law of distribution, we obtain a real form of arrangement characteristic of goods in a rigid warehouse - curve D_3 in Fig. 1b, which is necessary for comparison of their indices.

For demonstration of the results interpreted in Fig. 1, the arrangement of goods in the cells of an AW is practically illustrated in Fig. 2.

For calculating the volume of a warehouse required for arrangement of N units of goods in a flexible warehouse, it is necessary to use the same initial data of the goods flow, as for calculation of the arrangement in a rigid warehouse. Since the discreteness index fully characterizes the conversion to a flexible warehouse, we obtain the following calculation formula of $V_N^{(m)}$:

$$V_N^{(m)} = N \sum_{i=1}^{n^{**}} \sum_{j=1}^{m} \frac{1}{i} \int_{a_*}^{a^*} V P_v(V)\,dV \ , \qquad (4)$$

where

$$a_* = \frac{V_{max}}{i+1} + (j-1)\frac{V_{max}}{mj(i+1)} \ , \quad a^* = \frac{V_{max}}{i+1} + j\frac{V_{max}}{mj(i+1)} \ , \quad n^{**} = ent\,(q-1)m$$

It should be noted here, that it always holds true:

$$V_N^{(m_1)} \geqslant V_N^{(m_2)} \qquad \text{with} \quad m_1 < m_2,$$

and with the increase of the discreteness index m we have the following limit relation (which is also evident from the logical consideration):

$$V_N^{(\infty)} = \lim_{m \to \infty} V_N^{(m)} = N \int_{V_{min}}^{V_{max}} V P_v(V)\,dV$$

that is by determining the volume utilization factor in the absolutely flexible warehouse, $K_v = 1$, whence we obtain an important relation for numerical evaluation: $V_N^{(m)} \geqslant V_N^{(\infty)}$.

Before passing over to comparative characteristics relating to evaluation of advantages of conversion to a flexible arrangement of the warehouse, let us determine the volume utilization factor. For this aim, we determine the list of indices of the goods which will be used during the operation of an automatic warehouse as a component of a flexible manufacturing system.

One of the most important indices - dimensions of goods, have already been taken into account and are measured in practice by automatic measuring devices. Another basic parameter is the weight of goods. This is explained by the fact that, as a rule, during designing and using the racks of an automatic warehouse, stringent limitations are imposed on the weight of the goods kept in warehousing, which in many cases leads to limiting the filling of the automatic warehouse. For calculation of the said limitation we introduce an additional function $P_\gamma(\gamma, V)$, describing the distribution of density of probability of the specific volume weight against the volume of goods. The stochastic character of this function is determined by the fact, that even if the articles treated in the FMS are made of one type of material (mostly metal), the volume taken by the unit of mass of the articles depends on the container used, associates, pallets, as well as their structure for example, cavities in body parts). In this case, introducing the limitation of the loading capacity of the warehouse in the form

$$m_\Sigma \leqslant M$$

where m_Σ is the summary mass of the goods in warehousing, M is the total loading capacity of the warehouse, we can write the following equation for the amount of the goods in warehousing:

$$N \leqslant N_{max} = M \Big/ \int_{\gamma_{min}}^{\gamma_{max}} \int_{V_{min}}^{V_{max}} V P_v(V) P_\gamma(\gamma, V) d\gamma dV$$

where N_{max} is the maximum possible amount of the goods in warehousing, γ_{min}, γ_{max} are change limits of the specific weight by volume of goods.

In this way the limitation of the value of the utilization factor with full loading of an AW is determined by the following unequality:

$$K_v \leqslant \min(1, \overline{K}_v),$$

where

$$K_v = N_{max} \int_{V_{min}}^{V_{max}} V P_v(V) dV / V_0$$

V_0 is the total volume of the warehouse. Fig. 3 shows $\overline{K}_v$ versus the specific volume weight spread and loading capacity of the warehouse. Further, for evaluation of the utilization factor of a cell of the rigid warehouse, we introduce the value R_s equal to the relation of the difference of areas located under the curves D_2 and D_3 in Fig. 1b (cross-hatched) to the total area under the curve D_2. This value can be used as a generalized characteristic of goods in the FMS (for example, for the FMS oriented to assembly, R_s has the value in the order of 0.4, and for the FMS of body parts, about 0.2-0.25 and so on). For evaluation of its value one can use the equation for the value of R_s for the case of optimum utilization of the volume of the cells:

$$R_s = 1 - \sum_{i=1}^{n^*} \frac{1}{i+1} \Big/ \ell n \left(\frac{V_{max}}{V_{min}} \right)$$

where n^* is the maximum amount of goods in 1 cell of a rigid warehouse. Now we have all the data required for the calculation of the value ΔK_v, that is the increment of the utilization factor of the AW volume, when changing over to the flexible organization of its structure with rearrangeable dimensions of the cells. From equations (3) and (4) we have:

$$\Delta K_v = \int_{V_{min}}^{V_{max}} V P_v(V) dV \left\{ \left[N \sum_{i=1}^{n^{**}} \sum_{j=1}^{m} \frac{1}{i} \int_{a_*}^{a^*} V P_v(V) dV \right] - \right.$$

$$\left. - \left[V_{max} \, ent \left(\frac{N}{n^*} \int_{V_{min}}^{\frac{V_{max}}{n^*}} P_v(V) dV \right) + \sum_{i=1}^{n^*-1} ent \left(\frac{N}{i} \int_{\frac{V_{max}}{i+1}}^{\frac{V_{max}}{i}} P_v(V) dV \right) \right]^{-1} \right\} \cdot 100\%$$

besides it can be noted that

$$\Delta K_v \geq \left\{ 1 - \frac{\int\limits_{V_{min}}^{V_{max}} V P_v(v) dV}{V_{max} \sum\limits_{i=1}^{p-1} ent\left[\frac{N}{i} \int\limits_{V_{i+1}}^{V_i} P_v(v) dV \right]} \right\} 100\%$$

that means that the lower boundary is known, which even with incomplete information always allows for a preliminary evaluation of the benefit of the rearrangement. The value ΔK_v is the function of the law of distribution of volumes, discreteness index and characteristics of the load carrying capacity. In spite of the apparent complexity of the proposed equation, the results of the calculation with this formula can be represented rather graphically on the plots Fig. 4a, b of influence of the values R_s, q, and m on the increment ΔK_v in per cent.

The said dependencies are plotted for the normal law of distribution $P_v(V)$ and with determinantal $\gamma (V)$ = const. The dependences shown in Fig. 4a can be conveniently used when it is necessary to select the degree of flexibility which determines the constructional complexity of an AW. The plots in Fig. 4b are more convenient for consideration of the problem of the functional efficiency of a flexible warehouse. From the curves shown in Fig. 4 it can be noted, that with the increase of m, the relative increment of the volume utilization factor becomes less and less considerable, while enlargement of the range of spread of volumes results, on the contrary, in a higher efficiency of the flexible rearrangement, which can be explained by realization of capabilities of a more rational utilization of the AW volume.
With warehouse loading limitations, the increment of the volume utilization factor ΔK_v becomes less considerable, because of saturation of the loading space and limited number of goods, however this does not mean, that higher flexibility is not rational in this case: the complex evaluation of the warehouse efficiency must include its performance considered below, which depends on the flexibility and has different values with one and the same K_v.

Thus the dependences obtained by equations (3) to (7) and plotted in Fig. 4 permit to describe the benefit, that can be obtained due to reduction of the warehouse volumes, which in a number of cases additionally provides for either their arrangement in a more convenient place nearer to the equipment, or installation of an additional robot-stacker, etc.
From the results obtained, it is possible to derive one more, practically important index of increment of the amount of goods simultaneously kept in the warehouse with rearrangeable sizes of the cells. It will allow to define probable directions of realization of potential capabilities of the equipment productivity, increase in output of production, and, which is most important, to enlarge the flexibility of the whole manufacturing process due to the possibility of mastering the production requiring more capacious AW in the process of manufacture in the FMS.
In order to calculate the amount of goods which can be stored in a rigid warehouse with the general volume V_o containing accordingly

$c_o = V_o/V_{max}$ of cells, one can use the results obtained above, whence it follows, that with sufficiently great capacity of the warehouse $N_o^{(r)}$, the general amount of goods in a rigid warehouse is equal to (taking into account the limitations to the AW loading):

$$N_0^{(r)} = min \left\{ \frac{V_0}{V_{min} \sum_{i=1}^{P-1} \frac{1}{i} \int_{V_{i+1}}^{V_i} P_v(v) dv} , \frac{M}{\int_{\gamma_{min}}^{\gamma_{max}} \int_{V_{min}}^{V_{max}} v P_v(v) P_\gamma(\gamma, v) d\gamma dv} \right\}$$

The amount of goods in a flexible warehouse of the same volume as the rigid one will be designated N_0 (m) and calculated as follows:

$$N_0^{(m)} = min \left\{ \frac{V_0}{\sum_{i=1}^{n^{**}} \sum_{j=1}^{m} \frac{1}{i} \int_{a_*}^{a^*} v P_v(v) dv} , \frac{M}{\int_{\gamma_{min}}^{\gamma_{max}} \int_{V_{min}}^{V_{max}} v P_\gamma(\gamma, v) P_v(v) d\gamma dv} \right\}$$

Since we are interested in the increment of the amount of goods in warehousing ΔN, we can find it from the following equation:

$$\Delta N = V_0 \left[V_{max} \sum_{i=1}^{P-1} \frac{1}{j} \int_{V_{i+1}}^{V_i} P_v(v) dv - \int_{V_{min}}^{V_{max}} v P_v(v) dv \right] / \left[V_{max} \int_{V_{min}}^{V_{max}} v P_v(v) dv \sum_{i=1}^{P-1} \frac{1}{i} \int_{V_{i+1}}^{V_i} P_v(v) dv \right] \quad (6)$$

The value ΔN is the function of the value V_0, laws of distribution of volumes and arrangement of goods. For calculations by the equations (6) we use the same data, which were used for estimation of the increment of the volume utilization factor, that is the normal distribution of goods volumes and the determinantal volumetric density. In this way we shall obtain a series of plots shown in Fig. 5a, b.
Now we know the results of calculations of the cells dimensions re-arrangement probability coefficient influence on the volume utilization factor and on the amount of goods in warehousing. However, as mentioned earlier, the said parameters do not give exhaustive information on the results of the flexible organization of the automatic warehouses, because, for example, the same goods, arranged differently, leave the values K_v and N unchanged, however the tempo of transfer of articles increases due to a more convenient arrangement, and therefore in addition to the information on K_v and N, it is necessary to calculate the influence of the possibility of cells dimensions flexible rearrangement on the performance of the automatic warehouses, which will be understood as the number of transfers of goods between the cells and the goods issue/reception point with the help of the robot-stacker.
We designate this value of performance as ν.
Let us evaluate this index for both rigid and flexible organization of a warehouse. As it is mostly used in practice and adopted for the calculation model in the present paper, the automatic warehouses have rectangular configuration with similar cells, whose number in vertical line is assumed equal to a and horizontally equal to b, then the total number of cells in a rigid warehouse N_0 (r) is equal to:

$$N_0^{(r)} = ab$$

Let us assume the average time of the robot-stacker displacement by one step vertically as t_a and horizontally to t_b (not taking into account in each case the start-up, braking and so on). Let us first calculate the influence of the location of the goods issue/reception point on the value of ν.
For this aim let us introduce the values a_n and b_n of coordinates of the goods issue/reception point relative to the cells of the warehouse.

Then T_Σ is the summary time of goods transfer from all the cells to that point (or the time of loading, assuming the travel speeds equal in one direction) and will be equal to:

$$T_\Sigma = bt_a \sum_{j=1}^{a_n} j + at_b \sum_{j=1}^{b_n} j + bt_a \sum_{j=1}^{a-a_n} j + at_b \sum_{j=1}^{b-b_n} j$$

Assuming the number of cells sufficiently great, one can find the most rational location of the goods issue/reception point:

$$\frac{\partial T_\Sigma}{\partial a_n} = 0 \quad ; \quad \frac{\partial T_\Sigma}{\partial b_n} = 0$$

whence we receive its coordinates:

$$a_n = \text{ent}\ (a/2), \quad b_n = \text{ent}\ (b/2)$$

In this case we have:

$$T_\Sigma = N_0^{(r)} \left[t_a \left(\frac{a}{2} + 1 \right) + t_b \left(\frac{b}{2} + 1 \right) \right]$$

Thus the performance of the rigid warehouse is equal to:

$$\nu^{(r)} = 2 / (bt_b + at_a)$$

In order to determine the maximum performance, let us find out the optimum configuration of the warehouse, determined by the expressions for its dimensions:

$$\bar{a} = \sqrt{Nt_b/t_a} \quad ; \quad \bar{b} = \sqrt{Nt_a/t_b} \tag{7}$$

with this, the performance is equal to:

$$\nu^{(r)} = 1 / \sqrt{Nt_a t_b}$$

For calculating the performance for the flexible organization, one can use the data on increase of the volume utilization factor. With this, the time of the robot-stacker displacement by one step in both horizontal and vertical directions will amount to the following accordingly:

$$t_a \sqrt{1 - \Delta K_v} \quad ; \quad t_b \sqrt{1 - \Delta K_v}$$

so, the summary time of transfers will be decreased by the following value:

$$\Delta T = \frac{N}{4} \left[\left(\sqrt{\frac{Nt_a}{4t_b}} - 1 \right) t_b \left(1 - \sqrt{1 - \Delta K_v} \right) + \left(\sqrt{\frac{Nt_b}{4t_a}} - 1 \right) t_a \left(1 - \sqrt{1 - \Delta K_v} \right) \right]$$

This is the reserve of time during which the robot-stacker can perform operations of reception and issue of goods. Using the introduced probabilistic model of goods flow, one can determine the increase in performance of the warehouse. For this aim let us calculate N_A – the additional amount of goods beyond the volume of the flexible warehouse occupied by N units of goods on condition that N_A of goods is shifted to the issue/reception point for the time ΔT. The unknown quantity is expressed by the following equation:

$$N_A = N V_{max} \frac{\left(1 - \sqrt{1 - \Delta K_v} \right) / \sqrt[3]{1 - \Delta K_v} - \left(1 - \Delta K_v \right)}{\int_{V_{min}}^{V_{max}} V P_v(v)\, dV}$$

So, the performance of a rigid warehouse is equal to:

$$\nu^{(m)} = \frac{N - N_A}{N\sqrt{N}\sqrt{t_a t_b}}$$

The increment $\Delta\nu$ as percentage is expressed by the following equation:

$$\Delta\nu\% = \frac{\nu^{(m)} - \nu^{(r)}}{\nu^{(r)}} \, 100\% \tag{8}$$

Calculations by this equation allow to plot the dependences shown in Fig. 6a, b.

So, the above calculations allow to determine the change of the basic AW parameters after conversion to their organization with rearrangeable dimensions of the cells.

The results obtained concern only the functioning of the warehousing zone proper and can be represented as algorithms combining concrete characteristics of goods flows (including quantitative and qualitative indices of the goods), structural characteristics of AW, as well as the data of the robot-stacker, but only those, which were necessary during the calculations and taken in the most generalized form. For development of a flexible automatic warehousing system, it is necessary to determine the interrelation between the said two groups of parameters, with addition of dynamical and structural characteristics of the robot-stacker, which will enable their tying up together as a unit. This is necessary for the reason that in a FMS interaction with an AW occurs practically always through the robot-stacker, which therefore is an important link in the production chain, and in our case its development should be performed together with development of the design of the warehouse with rearrangeable dimensions of the cells.

The review of modern AW of FMS, as well as additional research [(1)] proved, that in developing a flexible automatic warehousing system it is necessary to take into account the following structural and dynamical parameters of a robot-stacker: load carrying capacity, travel speed, as well as static and dynamic indices of positioning accuracy. Since the first two parameters depend mainly on the power supply of the actuating devices and are directly connected with them, the research of the accuracy indices should be the major key for solving the problem. The static accuracy Δ_{ST} is understood here as an index determined by the summary error of positioning depending on both the capabilities of the control system and errors of sensors, as well as the quality of manufacture of mechanical parts (gaps), deformations of structural elements under the action of different loads, etc. The dynamic accuracy Δ_{mov} gives the evaluation of the time of delay of the robot-stacker till the moment, when the oscillation amplitude of the load carrying device attains the value below the permissible one. Thus, the interrelation of parameters of the warehousing zone and those of the robot-stacker can be defined as follows: the structural design of the elements of variable size cells, as well as requirements to warehousing of goods dictate the predetermined accuracy of positioning, while the performance of the warehouse determines the permissible time for correction of the dynamic error. The same groups of parameters exert influence in the reverse direction, which permits their joint development. For theoretical description of the dependences mentioned, let us introduce into our consideration a mathematical model of the structure of the prevailing type of the robot-stacker, Fig. 7a, which represents a bearing column 1 travelling on rails 2 (upper and lower) along the warehousing rack 3. The column bears the lift 4 with vertical positioning mechanisms and control of loading forks 5 which effect handling of goods and elements of cells. Fig. 7b depicts the kinematic diagram showing the shifts x, h, required for positioning as well as the main sources of oscillations in the given device, caused by the presence of elasticity and influencing the dynamic error: twisting around the axis of the stacker column - α and shift of the load carrying fork - β.

197

After setting-up the expressions for kinematic and potential energy, and basing on them, Lagrange equations of II degree, the following system of differential equations is obtained for calculation of the robot-stacker dynamics:

$$(m_к+m_n+m_в+m_r)\ddot{x}+m_n(\dot{\alpha}\tau_1\cos\alpha-\dot{\alpha}^2\tau_1\sin\alpha)+(m_в+m_r)\left[(\ddot{\alpha}\tau_2+\ddot{\beta})\cos\alpha-\dot{\alpha}(\dot{\alpha}\tau_2+\dot{\beta})\cos\alpha\right]=Q_x$$

$$(m_n+m_в+m_r)\ddot{h}+m_n+m_в+m_r=Q_h$$

$$(m_n+m_в+m_r)\ddot{x}+(J_в+J_r+J_n)\ddot{\alpha}+m_r\left[(\ddot{\alpha}\tau_1\cos\alpha-\dot{\alpha}^2\tau_1\sin\alpha)\tau_1\cos\alpha-(\ddot{x}+\dot{\alpha}\tau_1\cos\alpha)\tau_1\dot{\alpha}\sin\alpha+\ddot{\alpha}\tau_1^2\sin^2\alpha+\right.$$
$$\left.+\dot{\alpha}\tau_1^2\sin2\alpha\right]+(m_в+m_r)\left\{\left[(\ddot{\alpha}\tau_2+\ddot{\beta}(\cos\alpha-\dot{\alpha}(\dot{\alpha}\tau_2+\dot{\beta})\sin\alpha\right]\tau_2\cos\alpha-\tau_2\left[\ddot{x}+(\dot{\alpha}\tau_2+\dot{\beta}\cos\alpha)\dot{\alpha}\sin\alpha\right]+\right.$$
$$\left.+\left[(\dot{\alpha}\tau_2+\ddot{\beta})\sin\alpha+(\dot{\alpha}\tau_2+\dot{\beta})\dot{\alpha}\cos\alpha\right]\tau_2\sin\alpha+(\dot{\alpha}\tau_2+\dot{\beta})\tau_2\dot{\alpha}\sin2\alpha/2\right\}+m_n\left\{(\ddot{x}+\dot{\alpha}^2\cos\alpha)\dot{\alpha}\tau_1\sin\alpha-\right.$$
$$\left.-\dot{\alpha}\tau_1\sin2\alpha/2+\left[\ddot{x}+(\dot{\alpha}\tau_2+\dot{\beta})\cos\alpha\right](\dot{\alpha}\tau_2+\dot{\beta})\sin\alpha-(\dot{\alpha}\tau_2+\dot{\beta}^2)\sin2\alpha/2\right\}+к_\alpha\alpha=0$$

$$(m_в+m_r)\left\{\left[\ddot{x}+(\ddot{\alpha}\tau_2+\ddot{\beta})\cos\alpha-(\dot{\alpha}\tau_2+\dot{\beta})\dot{\alpha}\sin\alpha\right]\cos\alpha+\left[\ddot{x}+(\dot{\alpha}\tau_2+\dot{\beta}^2)\cos\alpha\right]\dot{\alpha}\sin\alpha+\right.$$
$$\left.+(\dot{\alpha}\tau_2+\ddot{\beta})\sin^2\alpha+(\dot{\alpha}\tau_2+\dot{\beta})\dot{\alpha}\sin2\alpha\right\}+к_\beta\beta=0$$

where: m_k, m_n, $m_в$, m_r are accordingly the mass of the stacker colums, lift, load fork, load; r_1, r_2 are distances from the lift axis to the centres of mass of the lift and the load fork with a load; I_n, $I_в$, I_r are the central moments of inertia of the lift, load fork and load; k_α and k_β are rigidity coefficients corresponding to elastic shifts of α and β; Q_x and Q_h are the generalized forces of action of the actuating devices for transfers x and h.

The above system permits to calculate all the required dynamic parameters of the robot-stacker and determines the interdependence of positioning modes and the required dynamic error set-up time depending on the design characteristics of the robot-stacker. The required maximum permissible error of positioning equal to $\Delta_{ST} + \Delta_{mov}$ can be obtained from consideration of the load placing processes and transfer of the elements of cells ensuring change of their dimensions. The equations and basic calculation method for these processes are considered in paper [(2)], and for our purposes we shall use the most essential result, that is the influence of the dimensionless coefficients k_α and k_β on the summary relative time τ of an operating cycle including the duration of approach to the desired cell, time delay for damping of oscillations and load handling. The relative and absolute parameters relation has the following form:

$$\tau=tH_m/V_m \quad ; \quad k_\alpha=к_\alpha\tau_1/J_\Sigma V_m \quad ; \quad k_\beta=к_\beta\tau_2/P_r$$

where: H_m is the lifting height of the robot-stacker,

V_m is the medium travel speed,

I_Σ is the summary reduced moment of the inertia,

t is time,

P_r is the goods weight.

The generalized relationship between the reduced values is shown in Fig. 8, whence it follows, that there exists a zone of the most rational design parameters of the robot-stacker, determined by the minimum value of τ.

So, the above results permit to determine the interrelation between the main elements of the rigid automatic warehousing system: the warehousing zone and the robot-stacker serving this zone.

Basing on the AW volume utilization factor, which is directly connected with the design of the cells, as well as on the performance determined by the robot-stacker characteristics, one can derive a criterion for selection of the flexibility degree, ensuring the maximum obtainable economical effect. It is necessary to take into account, that in each

198

particular case the economical calculation is performed with due allowance for the specific features of the automatic warehouse under consideration, which is determined by the selection of the relevant efficiency criterion.
For illustration, as the simplest criterion, we use the wide-spread linear functional of the type

$$E = \sum_i b_i y_i$$

where b_i are the weight coefficients; y_i are the characteristic of the AW, which can be taken from the theoretical expressions obtained in the proposed paper: y_i is the volume utilization factor, y_2 is the performance, y_3^{-1} are the costs of manufacture of the cell elements and the robot-stacker which we adopt in mean root square dependence on the flexibility index and working cycle time accordingly. The use of the present (and any other) criterion requires a joint solution of the equations obtained in the proposed paper by numerical methods with the help of a computer. For search of the AW optimum (as per the criterion max E) data the method of the fastest descent was used.
The calculation was based on characteristics of the two most widely used automatic warehouses of the IHI (Japan) and NOKIA (Finland) companies including:

Warehouse parameters	Number of cells	Cell dimensions	Load carrying capacity	Speed of robot-stacker	Warehouse height
Value	(pc)	(mm)	(kg)	(m/min)	(m)
IHI	800	2000x 1000x 1500	400	20	20
NOKIA	240	1200x 600x 800	250	30	14.4

Adopted as characteristics of goods flows were the normal distributions of volumes and specific weights by volume with indices: q=2.5, ρ =3. As a result of the calculation, series of plots were obtained for the criterion E versus basic parameters, Fig. 9, where zones of optimum values of the discreteness index are shown (cross-hatched). Similar plots were obtained also for the warehouse of the NOKIA company.
The results obtained for the IHI warehouse allow to draw a conclusion, that in case of the limited load carrying capacity and less efficient volume utilization of the cells, a rather high index of discreteness (or the order of 8 to 10) is required, then the performance increases by 20-25%, volume utilization factor by 10-12%, which is equivalent to increase in the number of unitary goods by 22%; however, to ensure these capabilities it is necessary to finalize the design of the robot-stacker with the aim of raising its travel speed up to 35-40 meters per minute with simultaneous calculation of the rigidity of its construction within 100 to 200 kilograms per millimeter. With sufficient load carrying capacity and high utilization index of the cell volume, the characteristics of the IHI automatic warehouse are close to theoretically optimum values, that is the calculated m is close to unity.
Similar results are obtained for the warehouse of the NOKIA company, whence it follows that with the discreteness indices 3-5 the efficiency will rise by 20%, the volume utilization factor will rise by 8 to 10%, and the number of goods in warehousing at a time can be raised by 15 to 20%.
On the basis of the theoretical research and calculation, one can draw the following conclusions:

1. On the basis of research of the mathematical model of the automatic warehouse with rearrangeable dimensions of cells and served by a robot-stacker, theoretical methods are developed for calculation of basic indices of such a warehouse and characteristics of the robot-stacker.
2. The influence of the robot-stacker basic constructional data on the dynamics, as well as on the process of loading and rearrangement of the warehouse cells has been studied.
3. Calculation methods are obtained for the ranges of values of the basic parameters of a flexible automatic warehousing system ensuring its functioning with the maximum efficiency, and the practical calculation has been carried out.
4. Combination of theoretical results obtained with the economical calculation made separately for each particular warehouse will make it possible to develop the flexible automatic warehousing systems witn optimum parameters.

ILLUSTRATIONS:

1. Amount goods accomodated in a unitary volume.
   (a) in rigid warehouse, (b) in flexible warehouse
2. Arrangement of goods of different volume in AW cell.
3. Influence of load carrying capacity on limitation of the volume utilization factor in AW.
4. Volume utilization factor versus:
   (a) values m and q with the known R_s, (b) values q and R_s with the known m, dotted graphs relate to limitations corresponding to ideally flexible AW.
5. Plot of growth of the amount of goods in warehousing:
   (a) values are shown of bending points of characteristic with limitations on K_v,
   (b) direction is shown of plots displacement with increase of spread of volumetric weights.
6. Plot of automatic warehousing efficiency increment (discreteness index m is given in logarithmic scale),
   (a) influence affect of volumetric weight spread is shown
   (b) optimum parameters calculation zone is singled out
7. Illustrations to model of automatic warehouse operation
   (a) key diagram, (b) kinematic diagram.
8. Influence of design parameters on working cycle time of robot-stacker (plotted in relative units).
9. Efficiency criterion value as a function of discreteness index
   (a) case of sufficient load carrying capacity of IHI automatic warehouse,
   (b) case of limited load carrying capacity of IHI automatic warehouse.

    L i t e r a t u r e:
1. Koroljev V.A., Sergeev S.M., Pavlov V.A. Realisation of industrial tasks by means of pneumatic robots, 10 ISIR, 1980.

2. Koröljev V.A., Sergeev S.M., Azarov S.P. Point-to-point pneumatic Robots for assembly. 9-th ISIR, 1979.

3. Королёв В.А., Сергеев С.М. К вопросу о точности позиционирования промышленного робота при роботизации установочных и сборочных операций. Промышленные роботы и их применение. Ленинград. 1976.

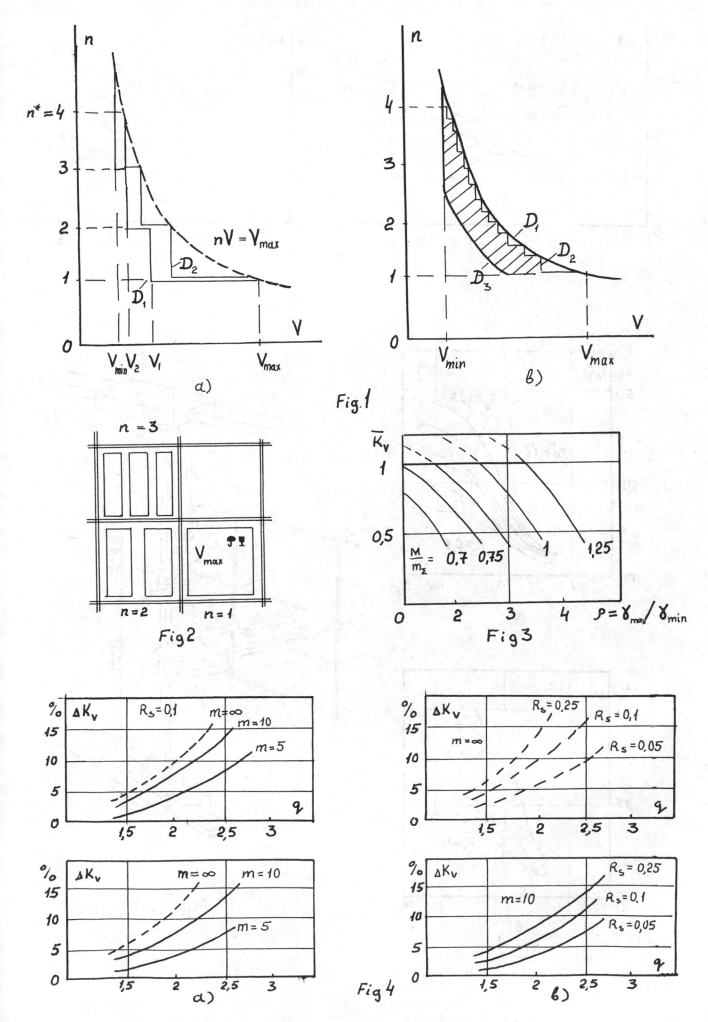

Fig.1

Fig 2

Fig 3

Fig 4

201

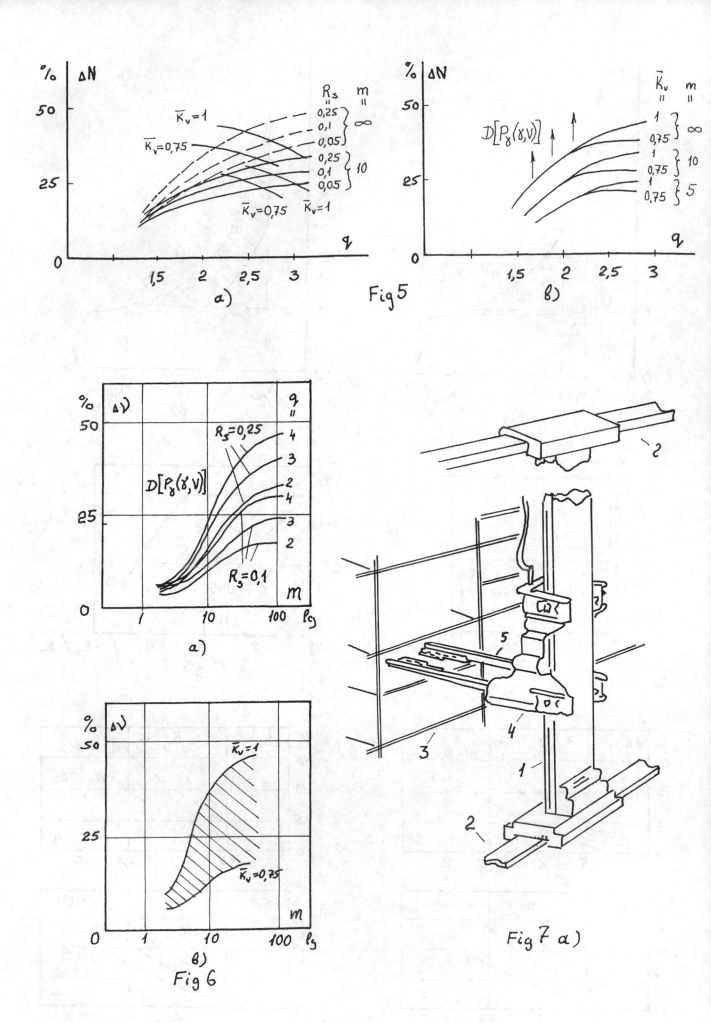

Fig 5

a)

в)

Fig 6

a)

в)

Fig 7 a)

202

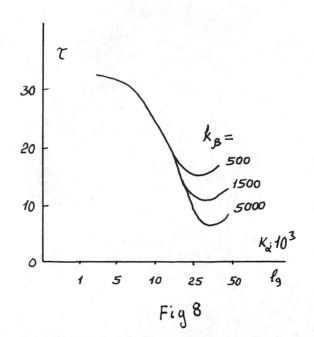

Fig 8

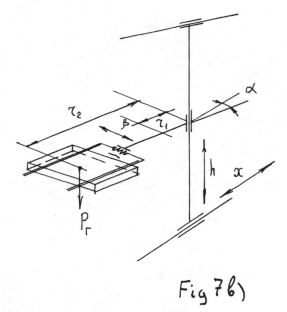

Fig 7 б)

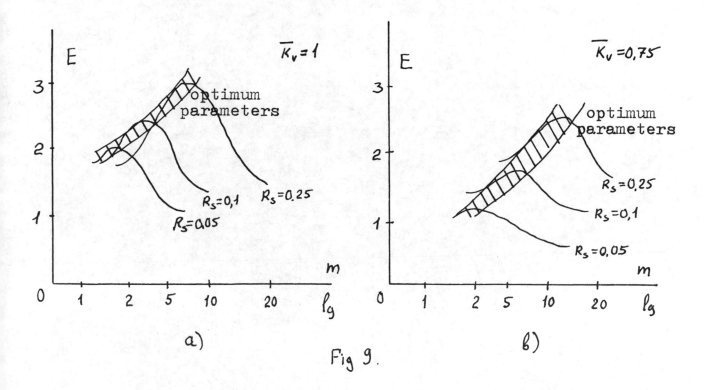

a)

б)

Fig 9.

A STUDY OF HUMAN FACTORS ASPECTS IN THE DESIGN OF UNIVERSAL CENTRE LATHES

H.S.Shan
Department of Mechanical &
Industrial Engineering
University of Roorkee
Roorkee-247 667 INDIA

ABSTRACT

The available designs of universal centre lathe are found lacking in considerations to safety and convenience of operator at work. The first phase of the present study consisted of a survey of a sample of industries where centre lathes of different makes, both recent and old, were being used. Operational features of these lathes were examined from human factors aspects. In the second phase, modifications were introduced in the design and location of some important controls and displays which improved the operator-machine relationship. The operators performance on the ergonomic lathe was then compared with that on the conventional lathe. The results indicated that the ergonomic lathe promises a higher level of operator's safety as well as machine productivity leading to low manufacturing costs.

INTRODUCTION

Universal centre lathes find most applications in job order shops where they are used for small or medium production work. An earlier study[1] had shown that although this is one of the oldest and basic machine tools, no major developments have taken place in its traditional design over the years. There are many features of this design which are inconvenient from operator's considerations. It was reported[2] that lathe operators were particularly liable to lower-back pain because they have to stand during the whole of the working day with the back bent and shoulders twisted in relation to the lower part of the body. There is, therefore, a need for the application of ergonomic principles to this production machine, but this relatively uncharming work is being neglected by research institutes, lathe machine manufacturers and user industries alike. The problem with the lathe manufacturers is that their companies are generally comparatively small and they are conscious of the conventional nature of their product (This is particularly true in the Indian industrial context). They would hesitate to go in for any major change because of the fear of non-acceptance of their product in the market if it is much different from the traditional design. In the recent past some enterprising manufacturers do have introduced new model of centre lathe into the market. These differ from the earlier ones in that they have better outlook, higher maximum speed, more number of speeds and are relatively more accurate and stable. But the aspect of convenience to operator appears to have been overlooked in these models too.

The aim of the present study was to improve the conventional design of centre lathe from the viewpoint of safety and convenience of operator at work. The work can be broadly divided into four parts :

(1) Collection of data on the relevant operational features of the available designs.

(2) Examination of various controls and displays from the human factors aspects in order to improve their design and layout.

(3) Incorporation of the improvements to develop what is called an 'ergonomic lathe'.

(4) Comparison of operator's performance on a conventional and the ergonomic lathe.

OPERATIONAL FEATURES OF AVAILABLE DESIGNS

A survey was conducted in five medium size job shops, two of which were famous for manufacturing high quality machined components. A wide variety of centre lathes-some old and some new models, all made in India were being used in these industries. These machines varied in regard to their motor capacity, bed length, and design and placement of controls. A representative selection of 10 lathes was made within the medium capacity range of 1-7.5 H.P. and bed lengths upto 2 m.

A summary of the relevant data of these lathes is given in Table 1.

Bed and Tool Post

Majority of the available models of centre lathe have low spindle height. When this is so, the tool post obscures the view of cutting action of tool from the standing operator. In order to monitor the cutting tool action visually, the operator bends forward and with his shoulders twisted in relation to the lower part of body, his posture is typically awkward (Fig.1). The number of times an operator bends forward and returns to normal standing posture in a working shift is quite large.

It was found that the tendency of operator to bend over the machine could be checked by raising the spindle axis height to his chest level and tilting the tool post so that the cutting tool in action can be clearly viewed from good standing posture. In the ergonomic model, the bed has been kept 30 inclined towards the operator and there is a provision for adjusting its height to suit the operator. Knee room has been provided so that operator can as well work in sitting posture when required.

Control for Engagement and Disengagement of Spindle Rotation

The starting/stopping of the main motor on most small and medium size lathes is generally achieved by operating push buttons or a toggle

Centre lathe capacity (H.P.)	1	1	2	2	3	3	5	5	7.5	7.5
Length of bed	140	100	180	130	165	200	200	180	200	200
Height of bed	86	85	85	90	85	80	82	82	75	78
Height of spindle axis:(1)*	105	110	100	108	100	110	105	110	110	105
from bed	19	25	15	18	20	25	23	28	35	27
Spindle drive	Belt	Gears	Belt	Belt	Belt	Belt&Gears	Gears	Gears	Gears	Gears
Number of speeds	3x2	8x2	3x2	4x2	4x2	4x3	6x2	8x2	8x2	8x2
Speed change control:Type	Crank/Lever	Crank/Lever	Crank/Lever	Lever	Crank/Lever	Crank/Lever	Lever	Lever	Lever	Lever
Location(1)	47,108	85	45,105	145	136	40,105	105,120	100,120	60,82	75,80
Location(2)	5,8	21	10,7	5	15	14,8	18,18	20,28	25,18	35,16
Operated by	Foot/Hand	Hand	Foot/Hand	Hand	Hand	Foot/Hand	Hand	Hand	Hand	Hand
Tool feed thumb wheel										
Axial feed: Size(dia)	14	15	17	16	18	18	20	25	25	20
Location (1)	76	50	72	77	72	70	70	68	65	65
Transverse feed: Size(dia)	9	10	12	10	12	15	10	12	10	10
Location (1)	88	88	88	94	90	83	85	85	80	82
Spindle stop-start control										
For stop: Type	Push button	Lever	Push button	Toggle	Push button	Toggle	Toggle	Push button	Push button	Push button
Location (1)	152	153	50	50	25	90	90	72	20	60
Location (2)	32	33	30	20	20	25	25	25	20	25
Operated by	Hand	Hand	Hand	Hand	Hand	Hand	Hand	Hand	Hand	Hand
For start: Type	Push button (Same as for stop)	Push button (Same as for stop)	Push button	Toggle	Push button	Toggle	Toggle	Push button	Push button	Push button
Location (1)	152	153	45	50	25	25	90	65	20	55
Location (2)	30	30	30	35	20	15	25	25	16	25
Operated by	Hand	Hand	Hand	Hand	Hand	Hand	Hand	Hand	Hand	Hand
Tailstock wheel dia	12	14	19	18	19	19	18	20	22	22

*Number in parenthesis refer to the key given below.

Key : (1) Measured from ground, vertically.
       (2) Measured from spindle nose towards left hand of operator, horizontally.

Table 1 : Operational Data of a Sample of Available Designs of Centre Lathes.

switch. It is interesting to note (Table 1) that the location of this control varies widely from one model to another model. On one machine the starting switch is located so low that the operator has to almost stoop to reach for it. From the operator's point of view how bad this location is can be imagined from this that in one working shift when a lathe is used in small batch production work the spindle rotation is engaged and disengaged 150 times on an average whilst in medium batch production it could be as much as 400 times.

The most suitable location of the motor starting switch is at or slightly above the operator's elbow level. But, safety and convenience of operator requires that the motor stop control should be a foot operated lever bar running throughout the bed length[3]. The ergonomic model has been fitted with a spring loaded hinged lever (A in Fig.2) which on being pressed, swings in the arc of foot movement of the standing operator.

Spindle Brake

The time taken by a running lathe spindle to come to a halt after the motor supply is cut off depends upon the speed at which it was running, friction in the spindle bearings, and whether or not any brake system is used. This time varies between 10-25 s in the lathes which have no spindle brake. More than 70 percent of the available lathes in the small and medium size range do not have any spindle brake. To cut short this ineffective time the rotating chuck must forcibly be slowed down. A common, though dangerous, practice followed by most operators to achieve this is to press the hand on the cylindrical surface of the rotating chuck immediately after the motor supply is switched off. This has sometimes caused finger/hand injuries.

The risk to operator can be eliminated and the cycle time reduced by providing an electromagnetic or band brake on the spindle. In the ergonomic lathe, a band brake has been used. The lever bar **A** (Fig.2) has been so designed that when pressed, the first half of its stroke cuts off the power supply to the motor while the later half operates the spindle brake.

Speed Change Control

In most of the centre lathes, the drive from the main motor is either completely through a set of gears or through belts and a set of gears. It requires the operator to operate two to three controls each time a change in spindle speed or tool feed is effected. Stop watch studies show that the desired speed and feed are obtained in these model in 4.0-5.8 s (mean 5.2 s) and 3.0-8.2 s (mean 4.8 s) respectively.

The speed and feed change times have been reduced by modifying the gear box design. It is now necessary to operate only one control for each, located at the convenient height, that is, just below the elbow height of the operator.

The possible ergonomic improvements in case of lathes which employ only belt drive system have been studied and reported elsewhere[3].

Display for Movement of Slides

For general machining work on conventional centre lathes, as much as four percent of operator time is spent[4] in examining or reading the movement of slides/spindle which are usually actuated by hand or star wheels. This movement is read from circular graduated scales fitted with these hand wheels (Fig.3). The accuracy of the reading of the movement is critical for good quality work. A man-machines study of the graduated scales being used on lathes in several metal working industries visited by the author has indicated that :

(i) The graduation marks on some scales are wide apart and operator has often to manipulate the reading. As the scale is on a circular surface the manipulation is difficult and time consuming.

(ii) The colour contrast between the graduation/vernier markings and the back-ground is often quite low. It is worse on older machines. This increases the chances of mistake in reading, particularly when the illumination level is low. Where frequency of reading is high, it may even cause eye strain.

(iii) There is no display to indicate the cumulative number of complete rotations of the handwheel. Thus, unless the operator is particularly careful, there is a possibility of making mistake causing the production of a defective part.

The circular graduated scales are, therefore, undesirable as they increase the time to read the value of movement, cause eye strain and production of defectives. These difficulties have been overcome in the ergonomic lathe by using digital indicator units (Fig.4) for the movement of main, cross and compound slides, and tailstock spindle.

Tailstock Controls

There are three main controls on a tailstock: for the spindle movement, spindle lock, and for clamping the tailstock body with the lathe bed. The design of this part of lathe is generally reasonable but on some lathes, access to the clamp bolt (Fig.5a) is not sufficient. The spanner can be turned only through a few degrees before having to re-locate the spanner. Lathes which have adequate space around the nut (Fig.5b) or in which clamping is achieved by operating a lever(Fig.5c&d) require lesser hand motions and therefore are better.

Operators' opinions were collected about the lever used for locking the spindle. Seventy percent of them indicated preference for the one which travels horizontally (Fig.5c), ten percent for the one which travels vertically (Fig.5d), and twenty percent showed indifference. In the ergonomic lathe, lever type tailstock controls (Fig.5c) have been used.

COMPARISON OF PERFORMANCE STUDIES

The performance of operator on the ergonomic lathe has been compared with that on conventional lathes. Some distinctive features are as under:

1. Raising the lathe bed and tilting it towards the operator enable the operator to have a good view of his work while standing in an easy, upright position. The operator has no tendency to bend on the machine. He can work just as well seated as standing on the lathe when sufficient knee room is provided under the lathe.

2. The coupling of brake control with the foot lever used for disengaging the power to spindle in the ergonomic lathe permits the spindle to be stopped quickly (Table 2). This has helped in reducing the ineffective cycle time, thus increasing the machine productivity. In addition, this is extremely useful in emergent situations like when the hands are engaged and the machine is required to be stopped instantaneously.

3. Modified design of gear box requires fewer controls to operate for effecting change in speed or feed in the case of ergonomic lathe. Further, the improvement in the design and layout of controls has reduced the speed/feed change times from 4.0-5.8 s (mean 5.2 s) and 3.0-8.2 s (mean 4.8 s) to 2.2-3.8 s (mean 3.5 s) and 1.5-3.4 s (mean 2.2 s) respectively.

4. The use of digital indicator handwheel units in place of the ones fitted with graduated scale has improved the flow of information between the operator and the lathe. As a result, reading times as well as possibility of making reading mistakes have been reduced[5]. Figure 6 shows a comparison of times to move the main slide through a given distance on otherwise two identical lathes.

CONCLUSIONS

It is possible to improve the operator efficiency and reduce the ineffective time of cycle by raising and tilting the lathe bed, improving the design and layout of the speed/feed change controls, using the spindle brake and by replacing graduated dials by digital indicators on the handwheels used for the movement of different lathe units. One can judge the effectiveness of these modifications in raising the machine and labour productivity from the fact that in one working shift when the lathe is used in small production work the spindle rotation is engaged and disengaged on an average 150 times, speed is changed 100 times and feed rate 140 times, while in larger batch production the figures are 400, 50 and 80 respectively.

ACKNOWLEDGEMENTS

The author is grateful to the Production Manager of Amardeep Industries, Ludhiana for help during carrying out a part of the study in his plant. He is also thankful to the subjects who voluntarily took part in the study.

REFERENCES

1. Corlett, E.N. "Research into the Ergonomics of Machine Control", Proceedings of the 5th International Machine Tool Design & Research Conference, Birmingham, pp 147-155 (Sept. 1965).

2. Floyd, W.F., and J.S.Ward, "Postures in Industry", International Journal of Production Research, Vol. 5, No.3, pp 213-224 (1966).

3. Shan, H.S., "A Study of Ergonomics of Centre Lathes", Proceedings of 3rd International Conference on Production Research, Kyoto Japan, pp 69-74 (July 1977).

4. Handa, S.K., "Time Study and Operation Motion Study in Some Metal Working Industries", M.E. Project Report, Department of Mechanical & Industrial Engineering, University of Roorkee (1977).

   Shan, H.S., "Activity Analysis of Machine Shop Operators", Unpublished Report, Department of Mechanical & Industrial Engineering, University of Roorkee (1978).

5. Shan, H.S., "Ergonomic Analysis of Machine Tool Design", Proceedings 4th International Conference on Production Engineering, Tokyo, Japan, pp 223-228, (1980).

Spindle speed range (rpm)	Average time to come to zero speed (s)	
	Conventional lathes (not fitted with spindle brake)	Ergonomic lathe (fitted with spindle brake)
31-100	10.5	1.5
101-200	11.5	1.5
201-300	13.0	1.8
301-400	15.0	1.8
401-500	17.5	2.0
501-600	18.5	2.0
601-700	19.5	2.0
701-800	20.0	2.0
801-900	20.5	2.5
901-1000	21.5	2.5
1001-1100	23.0	2.5
1101-1200	23.5	2.5

Table 2 : Stop watch study data of time to stop the lathe spindle.

Fig.1. A typical posture of a lathe operator when attempting to monitor the cutting tool action visually.

Fig.2. A foot operated lever used for both stopping the motor and applying the spindle brake.

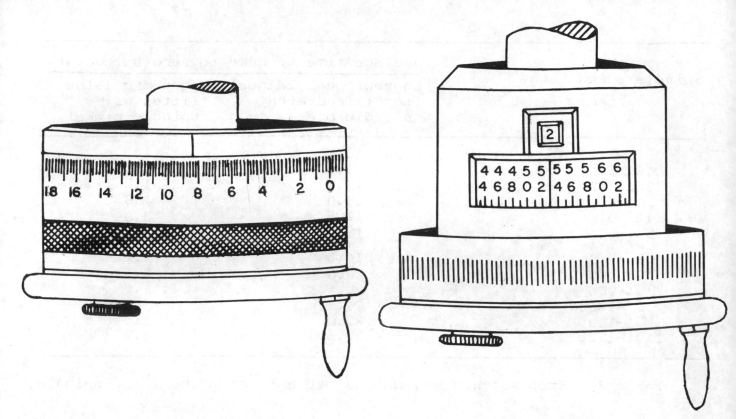

Fig.3. A graduated scale attached to the handwheel of conventional centre lathes.

Fig.4. A digital indicator attached to the handwheel for display of movement of slides etc. on the ergonomic lathe.

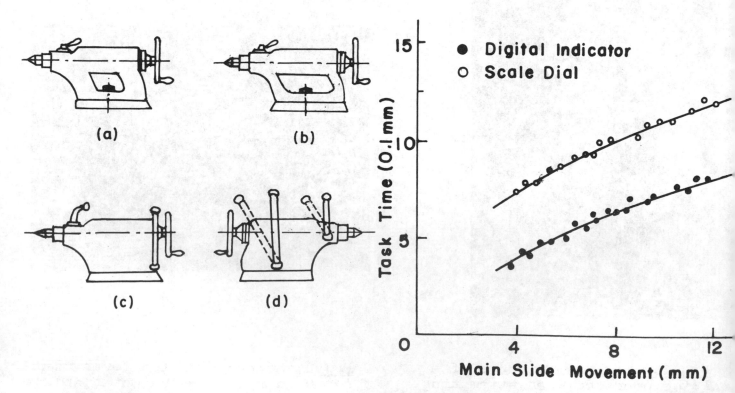

Fig.5. Tailstock controls.

Fig.6. Comparison of times for moving the main slide through a distance on the conventional and the ergonomic lathes.

ADVANCED WELDING TECHNIQUES

LASERS AND VISION SYSTEMS IN ROBOTIC APPLICATIONS

YEE MENG YEW, ROBOT MANAGER, ASEA SINGAPORE PTE LTD

ABSTRACT

The application of lasers has been extended to integrate in the robot systems to perform cutting, trimming, welding, hardening, inspection, etc. in the industries.

ASEA IRBS 101 Seam Finder, is the first available laser sensor uses to locate the actual position of a joint or workpiece, and simultaneously, select the correct parameter in thin sheets' welding.

ASEA Vision System provides automatic identification and localization as well as visual inspection of the workpiece. This is one of the latest development in the robotic field as the system can operate successfully despite variations in lighting intensity.

This paper will discuss the principles of above systems.

1. LASERS IN ROBOTIC APPLICATIONS

In robotic applications, laser is being transmitted to perform the desired tasks in various manners:-

1.1 Moving_workpiece_under_the_Laser_head

Some works have been conducted using an industrial robot to manipulate the workpiece under a static laser head. A typical application of this system is the inspection in a flexible manufacturing cell producing a family of similar components.

213

1.2 Laser head moving with Robot arm

The laser head/nozzle is mounted on the robot wrist and is positioned to a number of pre-programmed locations, where operation is performed in relation to the position of the robot. Three different systems are illustrated below:-

∘COBRA Beam Delivery System

COBRA system, a combination of beam guide and robotic assembly, is supplied by Ferranti, Dundee, Scotland. As shown in Annex I, the laser processing head is able to move freely over its entire working envelope with the wrist of ASEA IRB 6/2 robot. The system has a five degrees of freedom to maintain the laser beam normal to the workpiece.

The optical bench which houses 2 static goniometer mirrors directs the laser beam from laser head into the flexible beam guide. Subsequent mirrors within the flexible beam guide reflect the beam down to the processing head on the wrist of ASEA robot. Within the processing head, a lens focusses the beam. The focussed beam passes through an output nozzle with the required gas and hit direct onto the workpiece.

Both robot and laser operations are controlled by the robot's microprocessor unit.

The application areas for COBRA system are:-
- Cutting, trimming of stainless steel, titanium and nimonics sheets.
- Cutting of glass and kevlar fibre filled thermoplastics
- Trimming of forged components
- Laser welding
- Laser hardening

∘LASERFLEX™ 100 3D Laser Beam Delivery System

The system, as shown in Annex I, is supplied by the Industrial Laser Division of Spectra-Physics.

The beam enters the shoulder at a fixed inlet. Two rotating mirror blocks in the shoulder maintain the beam precisely in the center of the telescoping tube over the full range of motion. The wrist adds further flexibility to the system while maintaining the laser beam in perfect alignment through the exit aperature and lens/nozzle assembly. A bracket attached to the final block of the wrist allows ASEA IRB robot to control the exact location of the laser power.

The application areas for LASERFLEX™ 100 system are:

- contoured cutting and trimming of steels, aluminium, plastics, rubber, wood, ceramics, glass and cloth
- continuous welding and spot welding of three dimensional parts.

∘OPTOCATOR

Optocator, as shown in Annex II, is an optical range finder manufactured by the Swedish firm Selcom AB.

214

The optocator can measure the distance from the instrument to a surface within its measurement range. As per figure 1 below, the instrument consists of two parts, transmitter and receiver. The transmitter contains a laser diode which generates a laser beam. When this impinges on a surface, a diffuse reflection is obtained from the surface. The receiver functions as a camera and focusses a proportion of the reflection onto a diode sensitive to light. This diode registers where the light is focussed and generates a corresponding signal where is interpreted by the electronics to give a measurement of the distance between transmitter and object measured.

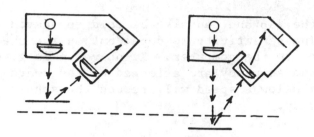

Fig. 1
Optocator
responds to
various height

ASEA is using the optocator in a robot inspection system. It is an optical measurement system which makes no contact with the object measured and can in no way damage or deform it. As the system consists one measuring sensor, the probability of error is reduced considerably. The inspection system is suitable for an "on-line" operation to detect when the measurements begin to depart from those specified and to provide a timely warning.

2. ASEA ADAPTIVE POSITIONING SYSTEM - IRBS 101 SEAM FINDER

2.1 In robotized arc welding, the robot leads the welding gun along a programmed path. For a satisfactory result, the joints must be in the correct position, only small deviations being acceptable. This can usually be obtained by means of fixtures and a certain degree of care when cutting and forming the sheet sections. With large workpieces, this is not possible or is very expensive.

The ASEA Adaptive Positioning System solves the problem of poor fit of the workpiece in robot welding and reduces the production cost in developing, manufacturing of fixtures and control of upstream processes. It enables the robot to position the welding gun appropriately at the joint before welding. Besides searching the joint, it also measures the gap at overlap joints which allows the robot automatically to adapt the welding parameters to the actual gap.

The adaptive positioning system as shown in Annex II, consists of an optocator, a micro-computer which evaluates the signal from optocator and the ASEA IRB AW control system with adaptive functions.

2.2 Principles of Searching

The measuring spot of optocator is created by a solid state low power laser. The optocator is mounted on the torch holder with the measuring spot around 20mm from the wire tip. When searching, the robot moves the optocator towards a surface or across an edge or a joint. The signal, the distance to the surface, is evaluated by the computer. When a certain distance has been reached or an edge/joint has been detected, a search stop signal will send to the robot.

In searching of height, the optocator is moved towards the surface and when it reaches the middle of the measuring range, a search stop signal is given.

When searching the side position of edges/overlap joints, the optocator is moved across the edge/joint. An edge/joint causes a step in the signal which is detected by the computer and a search stop signal is generated.

When searching the side position of fillet joints, the optocator is also moved across the joint. Search stop is generated when the distance after increasing starts to decrease.

The computer also calculates the gap and can also be told to search only positive or negative steps. The sensitivity is dependent on the sheet thickness which should be given to the computer. If no value is given, 1.0mm is assumed. The sensitivity and accuracy are affected by the search speed. Normal speed is 5-6 cm/sec. Too low a speed will reduce the sensitivity and too high will reduce the accuracy.

2.3 System Structure

The IRBS 101 Seam Finder is normally mounted on an ASEA IRB AW Robot. In order to get a good welding results, both the welding gun and the optocator are to be mounted in a fixed position which must not be changed during operation.

The best way to check the positions, as in normal robot welding, is to have a calibration point stored in the robot program. In this case, there shall have two calibration points. One for welding gun and one for the measuring spot.

The first time the welding gun is positioned perpendicular to the surface at a fixed point on the fixture, the robot position is stored. The position of the measuring spot is checked with the laser detector and is marked on the fixture. The height of the optocator is noted. When checking the mounting, the robot is moved to the programmed position and the side and height positions of both the welding gun and the optocator are checked.

The work of the adaptive positioning system is controlled by the Robot digital inputs and outputs. The laser has a very low power and the beam is focussed only close to the measuring range. However, one should avoid looking directly into the laser.

The optocator is protected against heat, fumes and spatter by an air stream. The clean air is supplied from a pump with a filter. The air-exhaust is prolonged by a hose in order to avoid disturbance of the shielding glass.

Besides this air protection, there are 2 shielding glasses which are easily replaced. When the optocator is vertical, the bottom can be removed by unscrewing 2 screws. The shielding glasses can then be replaced. In heavy welding, it is sufficient to change glasses every 8 hours. In light welding, the intervals between changes are much longer.

The above seam finder system can be used in most welding applications. It is however particularly appropriate for welding of thin sheets with many short welds where short cycle times are required. It is the first sensor system available for this type of industrial application.

3. ROBOT VISION SYSTEM

3.1 The use of industrial robots has become an established and effective way of increasing the degree of automation and flexibility of the manufacturing industry. It liberates humans from alienating and often hostile workplaces.

However, the supply of materials or workpieces to the robot station has often remained a monotonous, heavy, and in some cases a dangerous manual operation. Expensive dedicated mechanical arrangements, as an alternative, have masked the advantages of the robot, particularly in the case of small batches production or when a number of different versions are to be processed by the robot. Manufacturing automation is hampered by the lack of intelligent sensors. It is generally agreed that robots reach their full potential only when they are able to see what they are doing, evaluate the information thus obtained and act and respond to these changes in their environment.

ASEA introduces to the market industrial robots with integrated vision system. The system permits the robot itself to quickly identify different objects, accurately determine their positions and orientation and subsequently, command respective operation. It reduces the requirements for capital investment in peripheral equipment, the space it occupies and the associated engineering design work. Through the use of vision, it is possible to use low cost standard pallets, standard conveyors and grippers in the automatic manufacturing cell.

3.2 Terminology

◦Binary system
A system where an image is transformed into another image consisting of only two pixel values, zeroes or ones (pixel = picture element). There are a number of binary systems in the market, using different methods to achieve the binary image. Binary systems require very good lighting and they can only see the silhouette of an object.

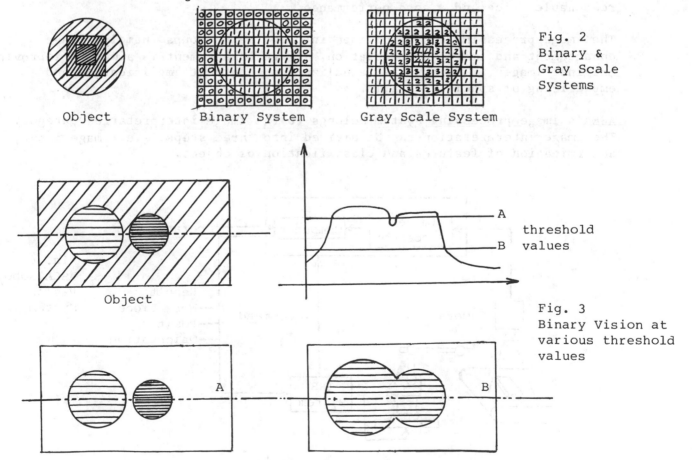

Object   Binary System   Gray Scale System

Fig. 2
Binary &
Gray Scale
Systems

Object

A
threshold
B values

Fig. 3
Binary Vision at
various threshold
values

A

B

217

○Gray Scale System
A system where the image is discriminated into a number of gray levels. This information is then used to extract the contours between areas with different gray levels.

○Level
Degree of grayness from black to white.

○Thresholding
All pixels at or above a preselected threshold gray level are assigned one. All other pixels are assigned zero.

○Adaptive Thresholding
If the background gray level varies due to non-uniform illumination and we know how it varies, we can use a variable threshold, where the threshold gray level depends on the position in the image. This is called Adaptive Thresholding.

3.3 Binary/Gray Scale Systems

The binary method does not give us full control over the process. If we have a low-contrast image, which usually is the case if we do not use black lighting, the exact value of threshold gray level can have a considerable effect on the form and size of the object. This makes it very sensitive to the threshold gray level.

The gray scale system can operate in normal industrial lighting and has the ability to register small differences in contrast.

3.4 Image Processing with Computers

Rapid advances in electronics, which have given us powerful microcomputers, better semiconductor memories and fast signal processing components, have now made it possible to manufacture image processing systems having both a reasonable price and a good performance.

The image processing can be divided into two main groups, namely, image enhancement and image interpretation. Image enhancement is aimed at improving original images, i.e. contrast equalization, contrast amplification and emphasizing of specific details.

ASEA's image processing system belongs to the image interpretation category. The image interpretation can be divided into three steps, i.e. image input, determination of features and classification of object.

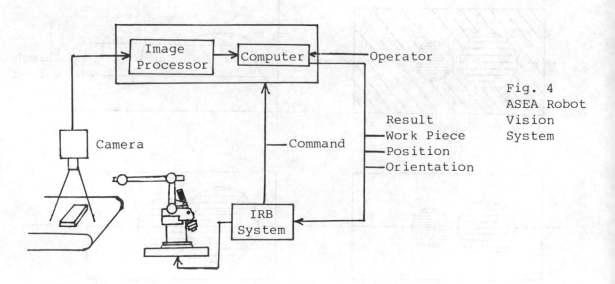

Fig. 4
ASEA Robot
Vision
System

218

An image processing system is comprised of an optical input unit, image processor and computer.

Image processing starts with the taking of a picture of the scene. This has to be converted to digital form to enable the computer to handle the information. A CCD (Charged Coupled Device) camera is used as input unit for the image processing system. This is a semiconductor camera having a two-dimensional image matrix. Features of a workpiece are scanned electronically and transmitted from the camera in the form of a video signal to the image processor. Each image element has one out of 64 (ASEA Vision System has 64 levels) uniformly allocated gray values between black and white. The image is then stored in a 65-Kbyte read/write memory (RWM). The subsequent filtering yields a gradient image. This information is stored in tabular form in the computer memory, where it is called scene table. If everything functions satisfactorily, this table contains sufficient information to identify with certainty which object is presented in the scene, where it is located and how it is orientated.

The purpose of the image processor is to create the necessary information for the scene table, that is to convert the video signal of the camera to digital information, filter the image, classify it, form contours and structure closed contours. For a normal image this takes around 300ms.

The computer has many different tasks: it checks, tests and controls different steps in the image processor. It computes the position and orientation of the object on the basis of information received from the image processor. It performs co-ordinate transformations and it looks after the communication with the robot computer. Finally, it is responsible for the entire man-machine communication.

3.5   Integrated Hardware

The fully integrated ASEA Robot Vision system for industrial environments comprises four main parts:-

. ASEA IRB Robot
. Camera - Up to four different cameras of the CCD type may be used. They have external synchronisation generated by the image processing system.
. Electronic part with memory and TV monitor integrated in the robot controller.
. Programming unit - this is the same unit as that used for the robot system itself.

The ASEA Robot Vision System uses the gray scale technique for image processing, which is suitable for normal industrial environments without any special auxiliary lighting. This technique enables the equipment to register small differences in contrast. The system is also capable of identifying varying forms on the surface of an object.

The image processing system is extremely compact and, with the exception of the cameras, is housed in the robot control cabinet. A 9 inches TV monitor is located next to the control panel in the cabinet.

The ASEA Robot Vision System permits the definition of 99 different objects viewed from one to six different directions. The storage capacity is highly dependent upon the complexity of the views. Normally, approximate 200 views may be stored at the same time. This means for example, that 99 objects with two views of each can be stored.

3.6 Programming

Programming is done according to a "memorising" procedure. Firstly, the object involved is placed beneath the camera. The system then processes the image and presents its contours on the TV monitor. The system is then "trained" a number of times in order to collect statistical data and to check if it really can recognise the object. Defining of the robot's grip point is also included in the Programming mode.

During the programming, it is always possible to correct mistakes, to modify data and to list stored data on the TV monitor. All image processing data can be stored in the floppy disc unit of the robot.

3.7 Consideration during Installation

When using the image processing system, more accurate results can be obtained by a proper selection of the lighting conditions, i.e. direct sunlight on the scene must be avoided.

There must be a contrast between the object and its background to ensure identification of the object. It is normally advisable to choose dark background material with a low luminance value in order to reduce the influence of shadows.

3.8 Applications

With the use of robot vision, it is possible to inspect hole location and number of holes in production parts to ensure correct assembly, for component verification or for detection of defects.

The other major use of robot vision is guidance and control. The vision system directs the robot based on what is seen and how this information is being evaluated and interpreted. Examples are part positioning, machining, processing, assembly and fastening.

The advantages of robot vision system are flexible installations with less project engineering and greatly reduced the need of expensive peripheral equipment.

A few of the most interesting installations are described below.

○Deburring - small batch production
Deburring of contactor housings is a typical job where batch size is often limited and the number of variants is large. In this particular installation, an ASEA robot with vision is installed and up to 12 different moulded plastic parts are deburred in one single robot station. In the work cycle, the parts are placed without any particular order on a conveyor. The vision system identifies the part and determines its location. The robot automatically select the right grippers and program, it picks up the part, remove the flash, finish drills holes and remove internal burr. Finally, it places the finished part on the output conveyor.

○Serving machines
A major car manufacturer feeds up to 22 different kinds of heavy shiny metal axes into a machine with ASEA IRB 60 robot with vision system.

○Loading welding fixtures and performing quality check
Another major car manufacturer supplies 10 different parts to a welding station on a conveyor belt. Here the robot identifies and grasps the objects using the vision system. Before welding is performed the vision system also ensures the correct position for all parts to be welded.

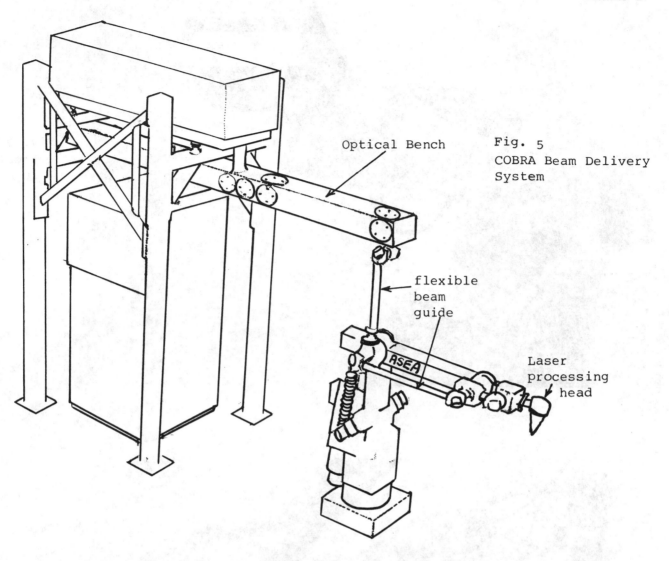

Optical Bench

flexible
beam
guide

Laser
processing
head

Fig. 5
COBRA Beam Delivery
System

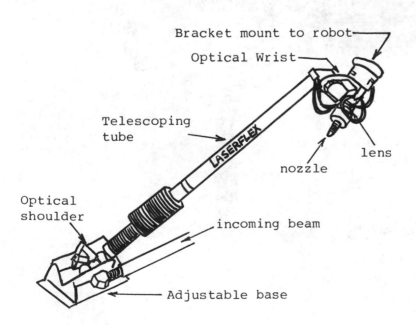

Bracket mount to robot

Optical Wrist

Telescoping
tube

nozzle

lens

Optical
shoulder

incoming beam

Adjustable base

Fig. 6
LASERFLEX™ 100
3D Laser Beam
Delivery System

Fig. 7
Optocator

Fig. 8
Seam Finder

A Robot Welding Cell for Small Batch Production

P S Monckton, R W Hawthorn, Wolverhampton Polytechnic, England
and
P M Hutton, Chubb & Sons Lock and Safe Co Ltd, England

This paper is concerned with the welding of
small batch security products using a robotic
cell. The cell has been developed as the result
of a collaborative project between the Polytechnic
and Chubb & Sons Lock and Safe Co. In the paper
consideration is given to the specification,
design, layout and operation of the cell including
safety precautions for the operators. The major
factors influencing the capacity and efficient
operation of the cell such as product mix,
component handling and program loading are also
included.

1 INTRODUCTION

The Chubb & Sons Lock and Safe Co Ltd is a company within Chubb Lock & Safe Ltd, a division of the Chubb Group of Companies. The Group activities cover the whole range of security products from locks, safes and strongrooms to fire detection and prevention equipment, burglar alarms, warden activities, and the manufacture of various items of safety equipment.

On the site at Wolverhampton, high security locks, safes, strongroom doors and fire resistant cabinets are manufactured. In particular, during the manufacture of safes and strongroom doors, a large proportion of the process route comprises welding. Presently, all the welding is performed manually, resulting in high costs. These costs can be reduced by automating the welding process.

Due to the nature of the work, the products are manufactured in relatively small batches, sometimes as low as 5. Therefore, to fully utilise an automatic welding system, a range of components will have to be welded and changes between components will have to be rapid. Only a robot provides the flexibility for automation in these conditions. There are many examples of robots being used successfully (1-3) for such applications. The main advantages being achieved are a reduction in weld cycle time and an improvement in weld quality.

This report describes the method used for selecting the components to be welded by a robot and the specification of the system best able to handle the identified product mix. A brief description of the operating procedure of the system is also included along with a discussion of the safety requirements.

2 PRODUCT MIX

2.1 Introduction

The product mix was obtained by gradually expanding the range of components to be welded by the robot until sufficient work load was generated to fully utilise the robots capacity at an assumed efficiency of 75% on a two shift operation.

2.2 Component Analysis

The welding time for a component was determined using the guidelines described by the Welding Institute (4). This consists of calculating the arc time of each component, which is a function of weld length and welding speed. It is then assumed that arc time provides 60% of the total cycle time, the remainder resulting from manipulation operations. The results obtained on a few components are shown in Table 1.

TABLE 1   COMPONENT CLASSIFICATION

Safe Type	Component	No of Welds	Weld Length (mm)	Weld Speed mm/mins	Arc Time (mins)	Cycle Time (mins)
	Door	16	119	10.0	11.9	19.8
A	Inner Body	12	40	10.0	4.0	6.7
	Outer Body	4	78	10.0	7.8	13.0

The production quantity of the components per annum was estimated from a survey of historical data at the company.

At this stage the accuracy and suitability of each component for automatic welding was assessed. It was established that in the majority of cases the components would have to be tacked prior to welding in order to maintain the required repeatability. Obviously, there were some instances where components were deemed unsuitable for welding by a robot and were omitted from the considerations.

2.3 Component Selection

Initially, machined components were included in the product mix due to their good repeatability. However, these components only utilised a small fraction of the cell capacity, and were small in relation to the other components. Therefore, it was felt that these components should be excluded from the final product mix.

The components finally selected for automatic welding covered three safe types and three family types. In this paper the three safe types are designated A, B and C; the three family types are:-

    1) Safe Doors

    2) Safe Inner Bodies

    3) Safe Outer Bodies

This resulted in a product mix containing up to fifty different components, since there is a variety of safe sizes within a safe type. Due to the variety of safes considered a wide range of component sizes and weights is present in this product mix. For instance, safe doors can range from 0.3 m x 0.5 m x 0.1m weighing 100 kg up to 2.0 m x 1.0 m x 0.1 m weighing 500 kg. Outer and inner bodies are of a somewhat greater depth (up to 0.8 m) but in general tend to be lighter in weight (50 kg). All these factors must be accounted for when deriving the cell specification and method of operation.

3 CELL SPECIFICATION

3.1 Introduction

The main aim of cell specification is to identify the most appropriate robot, component manipulator equipment, and cell layout for welding the range of components identified.

3.2 Robot Selection

The choice of robot will primarily influence the cell layout. As shown in Figure 1 there are two basic configurations:-

    1) track mounted

    2) fixed position

In general the smaller commercial robots are mounted on a track to increase their working envelope and thus enable them to handle large components.

There are a few commercially available robots that have sufficient reach to weld large components when fixed to the floor. Such an example is the Cincinnati T3/746 robot.

For the application at Chubb, a fixed position robot was chosen. The main reasons were ease of programming and simplicity.

3.3 Manipulators

There are two main types of manipulators used with robots:-

        1) Two axis manipulators

        2) Trunnion manipulators

For the Chubb application, it was decided that a trunnion would be needed for welding of doors. These components are shallow and have to be welded on both sides. A trunnion is necessary to enable the components to be turned over during the welding cycle.

In the case of the inner bodies, these are tall and have to be welded on the front and back. Hence a two axis manipulator is required to enable the component to be rotated through 180° in the horizontal plane during the welding cycle.

The outer bodies can be welded without any manipulation, since all the welds can be easily accessed by the robot. Therefore in this case a stationary table can be used.

Obviously, the cell requires three workstations in order to provide the required manipulation. Three stations are also necessary due to the small batch size of the components to be welded and hence the need for frequent fixture changes. In order to minimise the influence of the fixture change operations on the cell efficiency, the robot operates at two stations whilst the fixture is being changed at the third. The system finally selected is shown schematically in Figure 2.

4 FINANCIAL JUSTIFICATION

As each component range was added, the system specification necessarily varied. The greater the component range, the larger and more complex the cell became and the greater the capital cost. Therefore the final product mix was not only governed by the welding capacity of the cell but also by the financial cost of implementing it.

For the system to be purchased, it had to show a 25% discounted cash flow over its expected life. This was obtained through a 4:1 improvement in weld times and the better weld quality reducing both finishing and assembly times. Assistance was also received by way of a government grant.

5 CELL OPERATION

The majority of the components to be welded by the robot are manufactured from pressings. In order to obtain the required repeatability, the components are tacked in existing manual fixtures, and are then loaded into simplified fixtures in the cell ready for final welding by the robot. Work is proceeding to improve part preparation and design so that the components can be loaded directly into fixtures in the robot cell, thus realising the full potential of the system.

On average, the welding time of the components is 10 minutes. Therefore in order to ensure that the robot is fully utilised, the fixtures need to be loaded and unloaded quickly and efficiently. Two jib cranes have been included in the cell, and will be solely dedicated to loading and unloading components and fixtures.

As previously discussed, it is proposed to weld safe doors at the trunnion station, the outer bodies at the stationary table and the inner bodies at the two axis manipulator. This will result in a balanced loading at each station and enable complete safes to be welded if required. Obviously, with three welding stations and up to fifty different components, the sequencing of the components through the cell in order to obtain full utilisation can be complex. This problem has been examined by simulation and a set of rules for sequencing the batches through the cell in order to obtain optimum cell utilisation has been derived (5). The aim is eventually to create, from these rules, a set of algorithms that can be used on a micro computer. This will then enable the foreman to determine the optimum batch sequence at the beginning of each shift.

A separate computer is used to down load programs since the robot control computer can only handle a limited number of part programs. At any instant in time, the robot computer can only store data from one tape. This means that for efficient operation of the robot cell, the robot computer memory must be large enough to store the programs for all fifty components, since it is feasible that the cell could be required to weld any combination of components. This requirement could not be guaranteed without the need for duplicate programming. The software for the down loading facility, the ROBUK System, was supplied by the software house CAP (7). This enables any program that is stored on the systems disc to be loaded into the robot computer so that the robot can perform that task. The programs can be loaded into memory in any order thus giving the required flexibility. The ROBUK system also enables up to 4 robots to be controlled by the micro computer.

6   SAFETY PRECAUTIONS

Figure 2 shows that safety mats are used in the system to prevent the robot entering a workstation where the operator is working. After completion of the task, the operator must step off the mat and activate the area by pressing a nearby push button. This signals the robot that the area is ready. After finishing its present task the robot then automatically moves to that area and begins welding.

To prevent the operator approaching the robot, waist high fencing has been placed between the three stations. To prevent other personnel entering the loading and unloading areas, the cell is surrounded by fencing and anti-glare screens. The anti-glare screens also provide locations where the cell operation can be viewed without the need to enter the danger area. Signs have also been placed at various positions around the perimeter warning of the danger of robots and prohibiting entrance of unauthorized personnel into the cell.

The reasons for selecting this system is to ensure maximum safety with efficient operation of the cell. To achieve this aim, the load, unload and fixture change operations must be as rapid as possible. Due to the physical size and weight of the components and fixtures, it was felt that enclosed guarding around each station would inhibit these operations and reduce cell efficiency. Only 10% of accidents occur during normal operation (6). The majority of accidents occur during the programming of the robot, when the safety mats or the physical guarding would not be operative.

To minimise the occurence of such accidents, with this installation, all the operators have attended the manufacturers training course and are fully conversant with the dangers associated with robots. For every combination of components to be welded in the cell, a chart indicating the robot movements has been prepared and the appropriate one for that days production will be placed on the perimeter fencing. This will enable observers to recognise if the robot makes an unprogrammed movement and enable them to disable it using the emergency stop.

The entrance into the working envelope of the robot for maintenance and teaching purposes is controlled by a door with micro switches wired into the control cabinet of the robot. If the door is open the robot cannot be operated in automatic mode thus reducing the chance of injury to the programmer or maintenance personnel.

A fume extraction system has been included in the cell to minimise the amount of welding fumes in the shop atmosphere which are then exhausted to atmosphere. This system was chosen in preference to electro-static precipitators, since once in production, it would not be feasible to stop the cell operation in order to clean the filters, which would probably be a frequent operation.

In an attempt to minimise the heat loss from the factory, the fans have been wired in series with the safety mats. This means that when a safety mat has been activated the fan associated with that area is switched on.

7 CONCLUSIONS

A robot welding cell containing three stations has been introduced into the Safe Works at Chubb. This cell will handle up to fifty different components in small batch sizes, which means that the components must be loaded and unloaded efficiently in order to obtain maximum robot utilisation. To obtain this aim, the operator has been protected from accidents by using safety mats in the system.

8 REFERENCES

1   P Simpson
    "Arc Welding with a Cincinnati Milacron T3 Robot at Geeste Industrial Products"
    Presentation at Welding Institute Technical Group Meeting, Nov 1982

2   T Salt
    "Fabrication of Large Structures with Robots"
    7th Annual B.R.A Conference held in Cambridge on 14-16 May 1984

3   R Smith
    "Case Study in Robotic Welding"
    Metal Construction, April 1984, p 209

4   J Weston
    Welding Institute Report 103/1979

5   P S Monckton, R Hawthorn, R Jones
    "Simulation of a Robotic Welding Cell to be used in Small Batch Production"
    7th Annual B.R.A Conference held in Cambridge on 14-16 May 1984

6   "Safeguarding Industrial Robots, Part I: Basic Principles!"
    Machine Tool Trades Association, 1982

7   B Hunt
    "The Use of an Attached Microcomputer to Boost Robot Capabilities"
    CAP Reading Ltd, Publication 1983

8 ACKNOWLEDGEMENTS

The authors would like to thank the Polytechnic, Wolverhampton and Chubb & Sons Lock and Safe Co Ltd for their cooperation and for allowing them to use their facilities. Also the SERC/DoI Teaching Company Scheme for making available the funds.

The authors would like to record their particular appreciation to Les Briggs of Chubb & Sons for his valuable assistance on the practical aspects of safe manufacture.

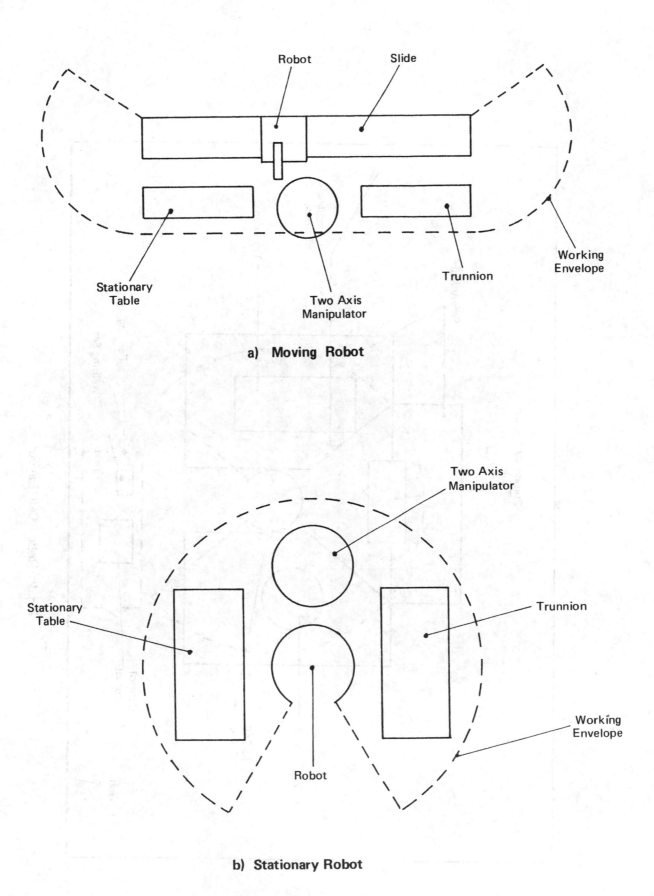

a) Moving Robot

b) Stationary Robot

Fig. 1 Possible Cell Layouts

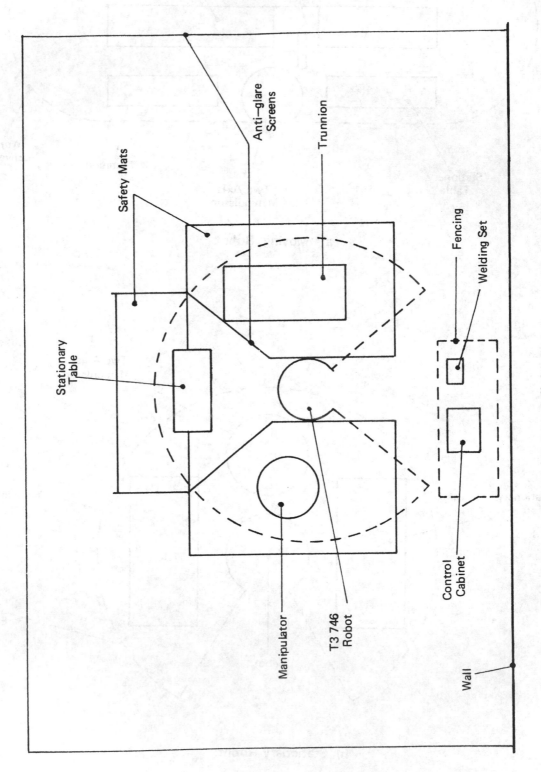

Fig. 2 Robot Cell Layout

Anti—glare Screens

Trunnion

Safety Mats

Stationary Table

Fencing

Welding Set

Manipulator

T3 746 Robot

Control Cabinet

Wall

230

ARC WELDING ROBOTS AND THEIR PRACTICAL APPLICATIONS

Gerhard Teubel, Dipl.-Ing.
Export Manager
Carl Cloos Schweisstechnik GmbH
Haiger, Federal Republic of Germany

1.   General Observations

In order to keep pace with increasingly keen competition, the vagaries of fashion, and the evolution of technology, manufacturers are obliged continuously to launch new products on to the market. First, the interval between the design and manufacturing stages must be reduced and, second it must be possible to use modern production machines for a wide range of applications. Multi-purpose machines have therefore to be built which are more flexible and easily adapted to the needs of technology and the economic situation. The robot satisfies all these criteria, which is why this technology has been the object of intense research and development for several years.

Nevertheless, when choosing an arc welding production system, one must not consider only the important features such as flexibility and adaptability. In each particular case it will be necessary to study the volume of production and any subsequent modifications during the amortization period evisaged. A machine, specific for a particular purpose, will often appear to be more advantageous, especially if the item to be welded permits simultaneous welding by several welding guns. In the automobile industry it appears clear that, with more than 800 units per day the specific purpose machine will be more profitable. But statistics reveal that even today the robot used for arc welding represents an important part of the market which, moreover, is showing a rapidly increasing trend.

2.   Description of a Welding Robot

2.1 Types of Robots

Basically, there are 3 types of robots, defined according to the type of movement by which they are actuated:

2.1.1.   Construction with cartesian coordinates.
In this case the movements of the 3 basic axes describe a cube situated on only one side of the robot.

2.1.2.   Construction with polar coordinates.
Here, the movements of the 3 axes are performed by a radial arm or an articulated arm. The working area is all around the robot.

2.1.3.   Construction with cylindrical coordinates.
In this case the linear axes are constructed according to the cartesian coordinate system and the circular axis according to the polar coordinate system. Because of this the working area is cylindrical and concentric with the circular axis.

The user must choose the model which is the most appropriate for the application for which he intends to use it. For arc welding operations, various models have been tested, year after year, and it is clear that the robot with an articulated arm, constructed with polar coordinates, gives the best practical results and provides the greatest possible diversity of application.

2.2   Axes and degrees of fredom

In order to describe the various movements by which the robot can be actuated, the terms "axis" and "degree of freedom" are used and refer to the system of cartesian coordinates.

The number of movements possible in a given system is called the degree of freedom "f" of this system. The mechanical devices causing these movements are called "axes". It is necessary, however, to take into account the fact that each axis does not necessarily correspond to a degree of freedom; in fact it often happens that two axes increase the geometric dimension of the movement in a certain direction without granting an additional possibility of movement with respect to the system of coordinates, which is why the degree of freedom is a nominal quantity (max. degree of freedom "f" = 6) whereas the axes of the system determine the size and configuration of the working area (principle axes) and often provide additional movement and orientation possibilities inside this area (articulated axes).

2.3   Robots with 5 or 6 axes?

Most robots on the market today possess 5 axes (3 principle axes and 2 articulated axes). For many applications, this concept is adequate, especially if the robot is completed by peripheral systems enabling the item to be welded always to be placed in the position suitable for welding.

Nevertheless, the robot with 6 axes is taking the lead, more and more, because the creation of software for circular interpolation, oscillation, seam tracking and access to difficult spots necessitates increased possibilities of movement by the robot. If this robot is them completed by peripheral instruments with numeric as well as freely programmable control, a system is obtained complying with the more stringent and more difficult requirements of arc welding. This system will then really deserve the name: "flexible production system".

Moreover, it is important to be able to use robots which can operate not only mounted on the floor but also suspended from the ceiling or in a horizontal position. These last two applications often offer great advantages since they enable the robot to be moved to a position above or to the side of items to be welded and thus to work under very favourable

232

conditions.

3.   Control design

All the control modules of a modern robot are arranged in the same control box so that there is no need to add separate interface components to it.

Generally speaking, the main parts of the processing unit, i.e. the central unit, the various processors, the memory and the interface modules occupy the top part of the control box. At the stage below, there is the input and data visualisation equipment and also the magnetic recorder. The servo amplifiers for the axis motors are all installed below.

The manual programming cabinet is used both for effecting the succession of operations and the welding parameters. The general configuration of the control system can be seen from the flow sheet; this has been envisaged so as to enable extension later, and the processing of data, eventually, on 10 axes.

4.   Characteristic of a Welding Robot

4.1 The user must be able to choose from several models. The choice will be made in terms of the robot's field of work and the geometric dimensions of the items to be welded.

4.2. Fidelity of repetition
The fidelity of repetition of the movement of the welding gun must be ±0.2 mm which presupposes great intrinsic stability of the robot's arm, the use of precision needle bearings and optical encoders having a resolution of 2056 signals per rotation of the driving motor. For the three basic axes, it is equally important that these encoders are mounted not level with the motor but immediately in front of the component carried by the corresponding axis.

4.3 Controlling the welding parameters
It must be possible for this control to be accomplished, starting with the programming console, and must include:
- rate of advance of the welding gun
- welding voltage
- current (wire speed)
- sweep amplitude
In addition, the first three parameters must be visibly displayed on the same console so as to facilitate control and, in particular, correction during the welding tests.

4.4 The writing of the program is carried out via the console keyboard and can be displayed on the screen. It must also be possible, however, to carry out this important operation at an independent programming station. Near the robot, the points in space will be simply programmed and the welding parameters optimised.

4.5 The control system must permit the programming of movements in the PTP (point-to-point) mode; the tool is guided towards the destination point as quickly as possible without taking account of the trajectory followed. The displacement mode is used in the approach phase and in disengagement from the workpiece.

In the CP (continuous path) mode, the trajectory followed is calculated. The CP mode is very important in welding since it enables simple geometric contours to be traced with reduced programming time. It is very easy to see, on the other hand, that in the "learning" mode, the robot can be moved along the cartesian axes X, Y, Z.

4.6 Circular interpolation

It is important that the system enables a correct angle of inclination of the gun to be programmed. For a circle, it is usually necessary to program three points at a maximum angular interval of 120°.

4.7 Simple programming of swept cords

The oscillation amplitude and frequency will be fixed during the writing of the program; their final control will be effected during welding tests.

4.8 Seam tracking

The most remarkable feature of a modern welding robot is the detection and tracking of the seam, i.e. in actual fact the correction of the real errors with respect to the path programmed.

Of all the systems at present on the market, it has been proved that the best is the one in which the arc itself provides the data necessary for the correction. The main advantage in this system is essentially due to the fact that no external component such as a magnetic deflector, mechanical detector or photoelectric cell is used whatsoever in the zone of action of the welding gun. Given the constant potential characteristic of the generator, the welding current intensity varies as a function of the length of the arc.

Example:

Soldering at 300 A, a variation of 1 mm in the arc length entails a difference of 6A in the current intensity. This value is sufficient to trigger correction pulses.

In order to track the seams, the welding gun must be given an oscillating motion.

For the normal geometry of an angular seam, we obtain:

```
                 L1 =  L3
    and thus     I1 =  I3
    and          L2 =  C (prefixed value)
```

In the event of error in the seam position the values above also vary and the control system receives correction pulses until the position defined can once again be attained.

Attention must be drawn to the fact, however, that this system also has limits of application.

In practice, one often has configurations and problems of very different characters, which is why this system must include several variants:

a)   in the case of a trajectory not geometrically defined, the system remains in action throughout the length of the weld run.

b)   the system operates only at the start of the weld run up to the bisector of the joint angle. The arc is then extinguished and the difference in the positions having been recorded by the computer, the welding gun is moved towards the starting point and makes the weld in the correct position, without sweeping. This system can be used where there is a possible movement of the steel sheets in

234

planes strictly parallel to themselves.

c)   for multi-pass welds, tracking the seam determines the new coordinates of the joint during the bottom pass soldering, and successively displaces the other passes by the correction values.

d)   if any movement of the whole welding assembly is involved, sensitive sensors mounted at the end of the arm and at the side of the welding gun are then used. These sensors can automatically relocate the correct position of the assembly and later on correct the postitions of all the weld runs to be carried out.

4.9   Permanent checking of the welding parameters

Permanent monitoring of the welding parameters is often indispensable for obtaining welds free from defects. The design certification of welding work by institutions such as the Bureau Veritas often includes such quality requirements. For certain very important welds, proof that all parameters have been maintained throughout the whole welding operation is obligatory. If the paramters exceed the acceptable tolerance limits, a visible or audible warning must be given. The machine must generally be stopped, and a computer printer must show the cause. For this purpose such a system has long been used for specific arc welding machines. When welding with a robot, the problem is complicated, however, since the robot often performs welding runs whose characteristics vary greatly, and consequently have completely different parameters. This is why a system must duly be created which permits individual monitoring of the tolerances of a large number of parameters.

The RPDU system satisfies these requirements. It permits the monitoring of 100 groups of paramters and gives a visual and audible warning if these tolerances are exceeded.The robot can be stopped and a print-out be provided; then the written proof of the situation and the reasons for the stoppage.

Since fluctuations of the order of a few milliseconds has no effect whatsoever on the welding result, it does not matter that the system whose cycle is 40 milliseconds gives a warning only after 5 cycles, i.e. 200 milliseconds. The parameter responsible for the error will be indicated as well as all the consequences relating to other parameters.

5.   The periphery of an arc welding robot

In a robot welding operation the welding equipment alone is not generally adequate:

-   sometimes, the position of the workpiece must change during the welding cycle, either for accessibility reasons, or as a function fo the welding position;

-   sometimes, the robot itself must change position in order to adapt its radius of action to the dimensions of the workpiece.

To satisfy the requirements, two solutions are possible:

1.   Use of rotating, possibly tilting tables, with or without a back edge.

2.   Use of longitudinal carriages whose path is a function of of the workpiece dimensions.

These two pieces of equipment may exist in two versions:

1. In the form of an indexing mechanism for all applications where:

   - the workpiece to be welded remains immobile during soldering (indexable rotating and tilting tables)

   - the robot does not neet to be moved during welding (indexable longitudinal carriage).

   In the second case, the running of the program is controlled by the input and output signals of the robot.

2. In the form of freely programmable exterior axes. In this version, the peripheral equipment can be moved during welding, thus enabling long or complex welding runs to be made without interruption of the arc. In this case, processing units for the supplementary axes are provided for in the control system. The movement of these axes is programmable and is monitored by incremental encoders just as the axes of the robot itself.

   It is therefore necessary for the user or his technical adviser to select from these basic versions the system most appropriate to the intended application, remembering to take into account other factors such as the weights of workpieces of the distance from their centre of gravity in the radial and polar directions.

6. The welding current generator

The final but not the least important consideration concerns the welding current generator. To enable the considerable advantages inherent in using a robot for welding operation to be fully exploited, the current source must have the following characteristics:

- it must be progressively controllable and programmable throughout the voltage and current intensity range (feeding of the wire) by means of the manual control box;

- for better initiation of the arc, a method of controlling the approach speed of the wire is necessary and also a reduction in the welding current in order to reduce the small beads which usually form at the end of the wire.

Recently, generators have appeared on the market which do not use thyristors for rectifying the welding current but a cascade of transistors.

The advantage of transistors is that they possess a very fast switching time which enables perfectly rectangular pulses to be obtained, as opposed to thyristor instruments which produce more "rounded" pulses. The rectangular pulse is characterised by having an intensity descent front at right angles which produces a very charply defined cut-off effect which detaches the welding bead, leaving no projection.

In addition, all the parameters of the pulsed arc can be controlled by potentiometer, i.e.:

- pulse frequency
- pulse width
- pulse height
- value of the basic current
- value of the wire speed

These control possibilites enable very regular and very precise transfer of the molten metal in the fusion bath which is the most appropriate for the application. The advantages are many, and notable ones include:

- welding without projections, thus eliminating grinding work and greatly reducing blocking by detached fragments (gaz nozzle, tube contact).

- a more stable arc, giving an excellent view of the weld run, and more compact deposition.

- possibility of controlling the heat application as a result of arc pulses, and the utilisation of wire of larger diameter but less costly to purchase.

7. Practical applications

7.1 The welding of steel shells

The automatic welding of the corners of steel shells always poses a problem due to the inevitable dimensional variations. Precisely for this case, an interesting solution has been found which can be adapted to other applications.

A corner of the workpiece is positioned against two fixed stops, and two robots simultaneously weld the opposite two corners.

In order to resolve the problem of the tolerances of the shell, the arm of each of the robots was fitted with sensitive sensors. These instruments detect the actual position of the workpiece to be welded and correct the position of the welding gun with respect to its programmed trajectory. The workpieces are mounted on an indexable panel, arranged horizontally, which enables them to be loaded and unloaded during the welding operation.

7.2 Welding railway wagon components

7.2.1 Suspension stand
A positioner may occupy 6 indexable positions in its axis of inclination whereas its axis of rotation is freely programmable by the robot software. The locking tool is actuated by a hydropneumatic system, and cooling is assured by a closed circuit system.

At the start of the welding cycle, a few tack welds are carried out and the interior final seam is also produced. During this operation, the movements of the welding gun and the rotating table have been programmed in such a way that the weld seam is always made in the chute position. After each welding cycle, the welding gun is automatically cleaned. In a "concealed" time, the other table, meanwhile, is unloaded and reloaded.

7.2.2 Haulage hook
This relates to a haulage hook with a support. Because of the relatively large dimensions of the corner seams the weld beads are applied in the chute position. The table is inclined at an angle of 135°.

7.3 The welding of container panels
Robots are also used in the production of containers. The following is an example of a robotised welding installation which shows the advantage of being erected on an overhanging wall.

Movement of the 6 robots is assured by three carriages, actuated by a programmable linear movement. This arrangement permits their access to all three different assemblies. The eccentric position of the robots with respect to the linear carriages increases their sweep of the two sides of the carriage. A central processor controls the interaction of the three carriages and ensures perfect synchronization between the work program of the 6 robots.

Needless to say, all the robots are fitted with a seam tracker to enable the inevitable dimensional variations of these assemblies to be followed.

7.4 Welding with several welding guns

In some cases one can even fit several welding guns on the tool holder of a robot. In this example, reinforced metal sheets are produced, with sections in omega for landing strips and foundations of industrial bays.

As the distance between the sections is always identical, the robot can be programmed for the simultaneous welding of three points and thus significantly reduce the cycle time.

7.5 Welding of gear cases

Production of gear-cases for civil engineering machines has also been robotised. Two positioners are placed on both sides of the robot. Each of these positioners is designed with 4 axes and a permitted load of 1500 kgs to receive workpieces weighing 1250 kgs. By means of 4 indexable axes, the programmer can place each seam in the chute position, indispensable for enabling the welding of seams with a groove of 9 to 13 mm.

Some details of this operation are given below:

- Depth of seams: 9 to 13 mm (in two passes)
  - 1st pass: with seam tracking, 350 A
  - 2nd pass: without seam tracking, 420A
- Wire used: DIN SG 2, 1.6 mm
- Deposition of metal/unit: 12.5 kg

During the complete cycle the assembly is tack welded beforehand. After the first welding cycle the positioner turns towards the outside where the person in charge adds other components which, for accessibility reasons, were not tack welded initially. During this time, the robot welds on the second positioner.

This method of working in"concealed" time gives the robot a very high utilisation factor, and consequently an appreciable reduction in production time is achieved.

COMPUTER-AIDED INSPECTION
AND ASSEMBLY

"COMPUTER AIDED INSPECTION FOR PRECISION MANUFACTURING INDUSTRIES IN SINGAPORE"

L J YANG
Senior Lecturer
School of Mechanical & Production Engineering
Nanyang Technological Institute
Singapore

ABSTRACT

In the past, Singapore's gross domestic product was dominated by commercial trading. Today, manufacturing is an important activity which accounts for about one-fifth of our total GDP.

Since the government introduced its economic restructuring programmes in 1979, the manufacturing industries which face a very tight labour situation have emphased more on mechanisation and automation. About six hundred NC/CNC machines and three flexible manufacturing cells have now been installed.

Computer-aided inspection not only improves the inspection efficiency and precision, but also alleviates the manpower shortage. The introduction of NC/CNC machines has also required the development of new inspection techniques and equipment. Computer-aided programming station, microprocessor or computer controlled co-ordinate measuring machine and automatic feedback gauging system are some of the tools which can be used for improving the inspection efficiency and productivity.

To date there are about forty (40) co-ordinate measuring machines, about forty (40) computer aided programming stations and about a dozen CNC lathes and machining centres with automatic feedback gauging systems installed in Singapore.

1. INTRODUCTION

When Singapore first achieved self-rule twenty-five years ago, its gross domestic product was dominated by wholesale, retail and re-export trading which then comprised about a third of it.

In due course, trading came to be challenged by the strong showing of three other sectors:

o   Manufacturing
o   Transport and communications
o   Financial and business services

Today each of these three activities is responsible for about a fifth of our total GDP as can be seen from Table 1.

The industrial history of Singapore can actually be divided itno three phases:

o   Industrialisation in the early 60s was triggered off largely by the need to solve the unemployment problem at that time. The government policy then was to attract labour intensive operations with liberal tax benefits.

o   In the 70s, the emphasis was shifted from one of import substitution to that of export oriented industries. Encouragement was given for manufacturers to move into the world market. With its strategic location as an ideal regional distribution centre, Singapore quickly established itself in this new direction.

o   Industrial development entered its third phase when in 1979, the government introduced its economic restructuring programme aimed at improving productivity and raising the technological level of the manufacturing sector. Since there is a shortage of labour supply, more emphasis has to be placed on mechanisation and automation.

2. HISTORY OF MECHANISATION AND AUTOMATION

The earliest examples of mechanisation were the use of multi-spindles and power feeds in machine tools. Later, sequence-controlled machines (eg. cam operated automatics) which could produce large numbers of identical components faster than manual machines were used.

The next step forward was the introduction of transfer lines for the machining of engine components. In this system, a group of milling, drilling, reaming and tapping machines were arranged along a conveyor along which workpieces were moved from one workstation to another where different operations were performed. Machine tools have now been designed with integral loaders, feeders, inspection devices and unloaders. These, combined with improved conveyor systems and manupulators have formed the transfer line production systems in which it is possible for work to progress from raw stock to finish parts without being touched by hand.

By automating work handling, machine tool and information feedback systems, it has been possible to attain very high levels of efficiency in manufacturing. The method of obtaining this efficiency has resulted in very expensive systems, rigidly designed for the production of specific items and only of use in the mass production sectors of industry. These systems are of little use in batch manufacture.

One of the most significant attempts at applying the technique of automation into batch manufacture was the introduction of NC and CNC machine tools. These are automatic machine tools which do not rely on mechanical means for sequencing and positioning functions but are based on more versatile and sophisticated electronic devices or computers for programming.

The advantages of the NC or CNC machines over conventional equipment in batch manufacture include;

o The ability to produce components of consistent geometry and quality at high rates for long periods. The cost of floorspace and manpower is reduced since one NC may replace three or four conventional machines.

o The amount of work handling and ultimately the lead time of a component and the amount of work in progress is reduced. This is because the use of long control programmes and automatic tool changers make it possible to combine many conventional operations into one NC operation.

o Rather then relying on jigs and fixtures for geometrical information as do conventional machines, the NC machine obtains all the required information from the control programme.

o NC machines make design changes more easy and cheaper and therefore increase the flexibility of the manufacturing system. This is because the cost of modifying the control programme is considerably less than that of modifying jigs and fixtures.

The future trend in manufacturing however will be on the flexible manufacturing systems. A typical flexible manufacturing system may comprise of a group of NC machines with synchronised work cycles, robots, conveyors and load/unload equipment.

The advantages of flexible manufacturing system include:

o Adaptability to changes in product design,
o Reduction of work in progress, and
o Improvement of productivity through faster work flow, reduction of waiting time and automated routing.

NC machine tools were introduced into Singapore in early 1970s. Although there were about twenty machines in 1975 (1), to date there are about six hundred units of NC/CNC machine tools and three units of flexible manufacturing cells installed in Singapore industries.

Figure 1 shows the growth of NC/CNC machines in Singapore.

3. COMPUTER AIDED INSPECTION

The activities within a manufacturing system may be divided into five broad areas:

o Product design/Identification
o Production scheduling
o Production planning and control
o Actual manufacturing processes
o Inspection and quality control

Manufacturers face many challenges in the 1980s. A key challenge is to manufacture high quality cost competitive products while increasing the productivity and profitability of existing resources. Since the primary force for increasing productivity and profitability is to improve quality, inspection and quality control is a very important activity manufacturers cannot afford to overlook.

Computer systems are particularly appropriate in seven key areas of quality control. These are:

o Automatic testing and inspection
o Data accumulation
o Data reduction, analysis and reporting

o   Information retrieval
o   Real-time process control
o   Statistical analysis
o   Quality management-related techniques

Automatic inspection and testing are used to:

o   reduce costs,
o   improve precision,
o   shorten time intervals,
o   alleviate manpower shortage, and
o   avoid inspection monotony.

The advent of NC/CNC machines has also required the development of new inspection techniques.

First of all, since the machine is controlled by a tape, the validity of that tape is critical. One form of inspection control is through certifying the tape or other devices by either running the programmes on the actual machine or by computer simulation.

The computer simulation technique is a better method for the following reasons:

o   It does not take up precious time on the actual machine.
o   It does not damage the tools or machines should there be mistakes in the programme.
o   A hard copy of the tool paths generated by the computer can be kept as a record.

Computer-aided programming stations supplied by machine tool manufacturers usually have both the hardware and software for doing a programme as well as for verifying the programme.

More advanced control units supplied by some manufacturers in recent years have also incorporated interactive graphic MDI functions onto their machines. These can also be used for programming simple parts and for verifying the tool paths.

Inspection of NC parts used to be done by the conventional methods of using micrometers, comparators and limit gauges. Although for NC parts, generally only selected dimensions need to be checked, however for reasons of improving quality and productivity and because of a tight labour situation, automated inspection seems to be the best solution in the long run.

Industries should make use of microprocessor or computer controlled co-ordinate measuring machines which likewise use the NC principle. The basic type of co-ordinate measuring machine is manually operated. It has the software for most applications, including 3D measurements.

The advantages of microprocessor/computer controlled co-ordinate measuring machines are:

o   Reduction of workpiece set up time
o   Improvement of measuring accuracy
o   Reduction of inspection time and labour
o   Flexibility in application
o   Automatic data processing

The advanced type of co-ordinate measuring machine comes with CNC controls. Fully automatic part measurement is possible with this type of equipment. This should be able to provide the most economical solution for measurement of identical parts over long spans of time, from parts in small-batch production up to measurement of mass-produced parts on pallets during the third shift.

The more advanced CNC machine tools have had automatic gauging systems connected onto the machine's computer. The feedback gauging system which can do precision measuring and tool position compensation automatically have the following advantages.

o Precision measuring and tool position compensation are completely automated by NC commands, eliminating manual measuring operations and the possibility of human error.

o Improved productivity with reduced cycle time.

o The result of each measurement may be automatically recorded by a printer.

4. STATUS OF COMPUTER AIDED INSPECTION EQUIPMENT IN SINGAPORE

Computer-aided Programming Stations

There are about forty (40) computer aided programming stations in Singapore. Majority of these are used for programming wire-cut EDMs.

Co-ordinate Measuring Machines

Although the manual co-ordinate measuring machine (e.g. validator from Brown & Sharpe) was introduced into Singapore about ten years ago, the first CNC coordinate measuring machine (a Zeiss UMM 500) was installed in a precision manufacturing company in 1980.

To date, there are about 40 units of co-ordinate measuring machines with micro-processor or computer control installed in Singapore. The major manufacturers of these machines are Mitutoyo, TSK, Bendix, Brown & Sharpe, Ferranti, LK and Zeiss. The accuracy of these machiens ranges from 0.02 mm to 0.0002 mm. The more advanced measuring machines not only have the capabilities of doing automatic 3D inspection, but also have the advanced softwares for doing form measurement for various component parts e.g. cams, gears and parts with complicated profiles.

The users of co-ordinate measuring machines in Singapore are:

o	Precision Engineering Industries	30%
o	Mould manufacturers	18%
o	Computer and peripheral manufacturers	18%
o	Electrical Appliances manufacturers	18%
o	Ordnance manufacturers	11%
o	Educational Institutes	5%

Majority of these co-ordinate measuring machines were installed after 1981.

Automatic Feedback Gauging System

Automatic gauging systems using touch sensors connected to the computer of the CNC machines have been introduced into Singapore industries since 1983. To date, a total of more than a dozen units of CNC lathes and machining centres have these devices incorporated. No attempts have so far been made for using laser sensors for direct non-contact measurement of dimensions of parts being manufactured.

5. CONCLUDING REMARKS

Computer-aided inspection not only improves inspection efficiency and precision, but also alleviates manpower shortage. Since Singapore industries have now entered into the high technology era with a tight labour situation; and in line with the installation of more CNC machine tools and flexible manufacturing systems, more computer aided inspection equipment should be installed. This will improve not only the quality of our products, but also the productivity of the challenging manufacturing industries.

ACKNOWLEDGEMENT

The author would like to thank Miss Aminah for typing the manuscript.

REFERENCE

1. Yang L J. "NC Machine Tools in Singapore", NC Machine Tool Seminar, Singapore Manufacturers' Association/Society of Manufacturing Engineers, 1975.

TABLE 1

THE SINGAPORE ECONOMY PAST AND PRESENT

	1960	1983
Per capita gross domestic product (in 1983$)	$ 2,700	$12,900
Trade*	34%	24%
Financial and Business services*	12%	22%
Transport and communication*	14%	21%
Manufacturing*	13%	20%

*   as % of total GDP
    Statistics from The Straits Times, June 11 1984.

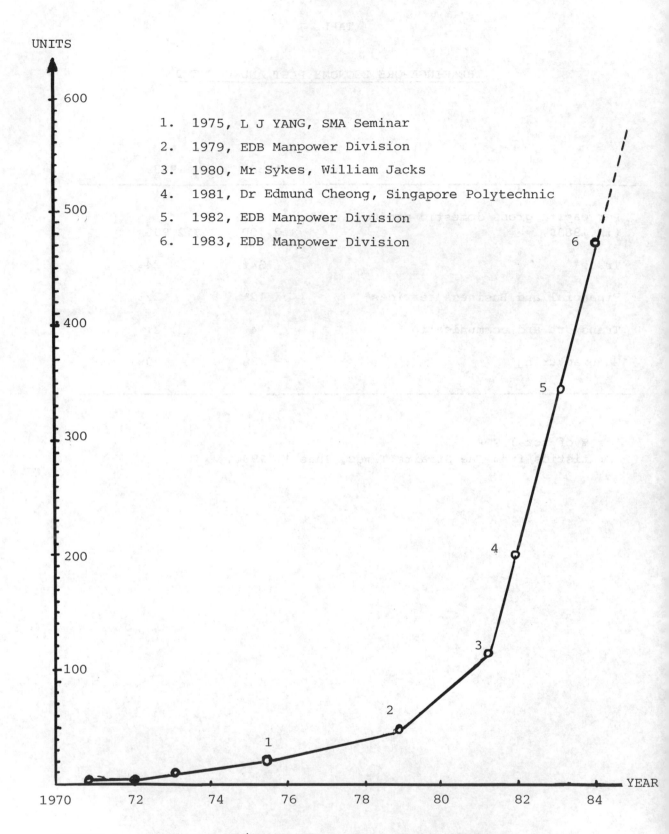

UNITS

1. 1975, L J YANG, SMA Seminar
2. 1979, EDB Manpower Division
3. 1980, Mr Sykes, William Jacks
4. 1981, Dr Edmund Cheong, Singapore Polytechnic
5. 1982, EDB Manpower Division
6. 1983, EDB Manpower Division

YEAR

FIGURE 1   GROWTH OF NC/CNC MACHINE TOOLS IN SINGAPORE INDUSTRIES

246

AN INTEGRATED ROBOT VISION AND TACTILE
AUTOMATIC INSPECTION SYSTEM

Harry Lowe
Ontario Robotics Centre, Canada

The aim of this project was to demonstrate how tactile and vision sensors could be used, on a robot, to provide a practical solution to an inspection task that was otherwise unreliable and labour intensive. The particular inspection task involved checking recently moulded automobile instrument panels for proper hole formation and the presence of flash around the edges of large openings.

1. INTRODUCTION

The aim of this project was to demonstrate how tactile and vision sensors could be used, on a robot, to provide a practical solution to an inspection task that was otherwise unreliable and labour intensive. The particular inspection task involved checking recently moulded automobile instrument panels for proper hole formation and the presence of flash around the edges of large openings. Originally, the work was undertaken as a feasibility study on behalf of a client. However, as the study progressed and evidence of feasibility became more apparent, the project requirement was changed to that of building a robotic workcell that would perform the above inspection task. This workcell was developed and tested at the Ontario Robotics Centre. Modifications have already been made to this design, to improve throughput capacity and a production unit should be implemented at the client's premises by early 1985.

2.0 THE INSPECTION PROBLEM

2.1 The Inspection Process

Currently, the plastic instrument panels are produced by injection moulding at a rate of forty parts per hour per injection machine. After allowing thirty minutes for cooling, parts are selected randomly from the production line for inspection. Inspection is performed manually at a rate of approximately 50 to 60 minutes per part.

The part selected is taken to an inspection room and placed on a jig, which is specially sculpted to correspond to the exact shape of the panel. In addition, the jig is designed so that white painted areas show through the moulded openings in the panel. These openings are specially shaped to receive the radio, cigarette lighter, air ducts, etc. and therefore have a critical size and shape. The painted areas correspond to the specified shape and size of the relevant openings. From an examination of these areas, it is intended that flash, or other irregularities in the opening, will be more easily detected by the inspector as a result of being highlighted by the white background.

Apart from the shape of these openings, the panel is also inspected for correct size of the openings and the presence and minimum depth of various blind holes. These holes are used for various screw and fastening devices which are inserted automatically later in the production cycle. It is, therefore, most important that these holes are present and properly formed. Should a fault in any of these holes be discovered, the parts are either reworked or scrapped. Should the fault be missed and not discovered until the screwing operation, the panel could be damaged, or ruined, and huge amounts of material and production time may be lost. At present, the blind holes are inspected manually using pin gauges. For some panels, up to 60 holes must be inspected in this way. The size of the large openings are checked by the inspector using various sets of calipers.

2.2 The Problems

There are three major problems associated with the above inspection methods. The method is: a) time-consuming, b) labour intensive, c) unreliable. All of these problems can be traced to a common factor: the use of a human operator to perform the inspection task.

The task is fairly complex in: amount of insertions, measurements and calculations performed. Therefore, it is unreasonable to expect the inspector to complete the work any sooner than he does and still maintain a reasonable degree of reliability. The time consuming nature of the inspection task directly affects the sampling frequency at which parts can be tested. If an inspector is checking the output of two machines, it would be possible to have up to 100 faulty parts produced by one of the machines before the fault could be detected (given a two hour sampling period per machine and a 30 minute cool down period).

The very nature of most inspection tasks has traditionally required that they be performed manually. Thus, when inspection is a requisite it presents an expensive fixed cost burden on any job.

The question of reliability is a well known problem in inspection. Who shall check the checkers? The very nature of industrial inspection tasks imply repetition and monotony. Both of these characteristics induce fatigue and lack of vigilance in human behaviour. The longer the inspection task, the greater the reduction in inspection reliability.

The correct sized gauging pin has to be used for each hole tested. If the wrong pin is inadvertantly selected, it is possible a batch of flawed parts will be allowed into the line. Also, when visually inspecting the large shaped openings, it is possible to overlook small areas of flash that may later interfere with the insertion of other components.

Since these types of error do occur, and their consequences are severe, it was decided to investigate the possibility of checking the instrument panels by means of an automatic inspection process.

3. INSPECTION SYSTEM REQUIREMENTS

It was decided at the commencement of this project that such an automatic inspection system must have the following design requirements:

a) inspection to be carried out off-line, in the inspection room.

b) equipment to be reprogrammable to inspect at least four different types of instrument panels.

c) inspection to require no human supervision.

d) inspection records to be maintained automatically with ability to determine trends should this be required.

e) be able to detect flash wider than 1/32" along edges of all specified openings (or greater than 2" in length, if not present along complete edge).

f) verify presence of all blind holes and check for minimum depth of hole.

g) provide a fast, simple method of fixturing that allows the panel to be inspected from both sides.

h) inspection time to be less than 30 minutes.

248

4. THE AUTOMATIC INSPECTION SYSTEM

4.1 Design Considerations

The following considerations were taken into account, in light of the above design requirements:

a) As the automatic inspection workcell was to be located in the inspection room, the maximum size of the workcell was determined by the available space. It was determined, therefore, that the workcell could not be greater than 16 feet x 16 feet and no higher than 10 feet.

b) The inspection system had to be able to perform a large number of insertion, or probing, operations at different angles of approach. Also, the inspection system had to be flexible enough to be easily changed over to inspect another type of panel. Because of the needs for manipulative manoeuverability coupled with flexibility of work task and scheduling, it was decided to use a robot as the basis of the automatic inspection workcell. However, in order to use a robot, the repeatability error of the robot had to be insignificant compared with the required measurement accuracy of the inspection system. As the system was required to make measurements to an accuracy of 1/32" (0.032"), the combined repeatability error of the robot and inspection fixture was thus required to be no greater than ±0.003".

c) As a robot was to be used to perform the inspection, part of the requirement for no human supervision had been answered. However, in order to have the complete inspection process unsupervised, it was required also, that the fixture automatically position and clamp the panel in a known location with a low repeatability error. The only requirement then made of the human operator was that he place the panel on the fixture and activate the inspection sequence. The operator then need return only when the system indicated completion of inspection.

d) The need to provide inspection records could be answered by using a separate control computer to analyze the inspection data and format the records. However, a small number of robots on the market also have this capability as part of their system software. These robots can be easily programmed, in a high level language, to print reports in any format required.

This requirement of the robot controller, coupled with the need to use a robot that can measure a panel 60" long and has a repeatability error of the order of ±0.002", severely limits the choice of robots that could be used for this task.

e) In the past, detection of flash around holes and openings has always been done visually by the inspector. Other methods may be considered, such as the use of tactile probes or pressure sensitive feeler gauges. These methods, however, would require constant monitoring, extremely accurate positioning and quite slow movement over the edges to be inspected. It was unreasonable to think that inspection time could be improved if this type of sensor gauge was used.

Machine vision was judged to be the most obvious type of sensor to use for this type of inspection. Machine vision has the advantage that it is non-contacting and therefore does not limit the movements of the robot. It also makes use of very compact sensors (miniature video cameras) that can be mounted on the end of the robot arm to provide close-up inspection from a wide range of viewing angles. Machine vision, also, is very fast when compared with other methods of performing this type of inspection.

f) The inspection requirement for the blind holes in the panel was to check that they were present and that they had a certain minimum depth. It was decided that both of these requirements could be answered by using a robot mounted probe to check for holes at each specified location on the panel. For this inspection, it is only required that the robot move the probe to a specified location and attempt to insert the probe into the hole to the specified minimum depth of the particular hole. Should the robot be unable to insert the probe to the full depth, this condition must be detected so that the operation can be aborted and failure date recorded.

A tactile, touch-trigger type of probe was chosen to perform this particular inspection operation. This type of probe is sensitive to deflections in any direction and provides a discrete I/O signal when activated. This signal could then be used to interrupt the insertion operation of the robot, causing the controller to record hole depth and position information.

Some consideration was given at the

beginning of the project to the use of other types of non-contacting sensors to detect presence and measure the depth of the blind holes. While some methods are available for locating the holes (eg. structured light machine vision systems), other methods, for determining the depth of these holes (eg. laser triangulation, ultrasonics), would have to be customized as they are not available in a suitable commercial form. All these methods, however, whether for hole detection or depth measurement, are far more complex than the touch-trigger probe. This type of probe requires no programming, no sophisticated maintenance, needs only a single discrete I/O channel, has more accuracy than needed, costs only a fraction of the other methods and is simple, robust and off-the-shelf. From these considerations, the use of a touch trigger probe was chosen as the most

suitable method of inspecting the blind holes.

g) As noted in section 4(c) above, in addition to the requirements of the fixture that it be simple and fast to operate, the fixture must also operate automatically to correctly position and clamp the panel for inspection. In order to combine the requirements of positioning and clamping, it was decided to use suction cups to pull the panel into place and clamp it in that position. It was necessary, therefore, to incorporate accurately positioned locating stops and supports into the frame of the jig, such that, when vacuum is applied to the cups, the panel is pulled against the stops and firmly held in position.

It was decided, also, to include an extra

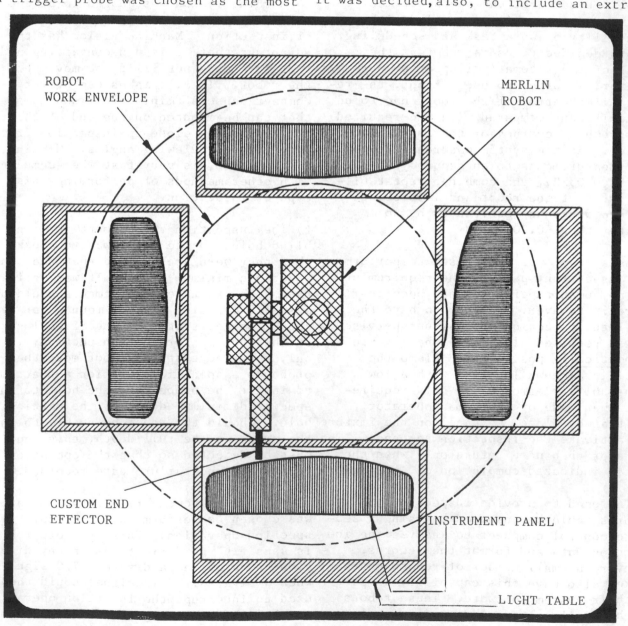

Fig.1 WORKCELL LAYOUT FOR PROPOSED AUTOMATIC INSPECTION SYSTEM

feature in the construction of the fixture. Some instrument panels have blind holes located on both sides of the panel. Consequently, it would be necessary for the robot to access both sides of the panel. Various methods of access were considered, including the use of variable-position probes and multi-probes mounted on the end of the robot. However, the final method decided upon,was to stay with a single fixed-position probe and have the fixture swivel about its major axis between two known and repeatable positions. As a final refinement, in order for the operation to be completely automatic, a double-acting pneumatic cylinder was to be used to swivel and clamp the fixture. Hydraulic dashpots would then be used to cushion the action of the fixture against the endstops.

h) The total inspection time of the system could only be determined from actual use. However, an obvious way of reducing robot movements, would be to perform the visual inspection and probing in one pass across the face of the panel, rather than two separate passes. Initially, however, until more was learned about the system, the two modes of inspection were performed on separate passes of the robot.

4.2 Workcell Operation

Workcell Layout:

The production version of the automatic inspection workcell will include four separate inspection fixtures located equidistant around a gantry mounted robot (see figures 1 and 2). The robot will be mounted upside-down from the gantry in order to conserve space in the workcell. For the test unit built at ORC, however, space was not a problem and the robot was floor mounted. Also, only one inspection fixture was used in this workcell as others would be redundant for the purpose of testing.

The workcell built at ORC comprised a two-position inspection fixture mounted on a backlight inspection table, a machine vision system and a six axis robot equipped with an end effector supporting the touch-trigger probe and a miniature video camera. The robot controller was mounted on the periphery of the workcell as was the vision system controller.

Sequence of Operations

The following sequence of operations

describes the action of the automatic inspection system:

a) Parts to be inspected are manually placed on the fixture in the correct orientation. The inspector hits the "START" button and the panel is automatically positioned and clamped to the fixture by the suction cups.

b) The Robot then moves to the right side of the I. P. and, using the Touch-Trigger probe, locates the part and registers its position. The Robot then moves to the other side of the part and registers the position of the left side. The absolute value of travel is taken as the parts length and compared to specified tolerances. If the part is within the specified parameters, it is accepted and inspection proceeds. Should the part be out of tolerance it will be rejected and the system will abort at this point.

c) Once the part has been accepted, the Robot receives a command to proceed with inspection. The Robot then moves in sequence, to the locations of each of the blind holes and checks for their presence (or absence). If the trigger probe detects a missing hole, a message is displayed on the CRT monitor stating which hole is missing. Since all blind holes have an identification number moulded into them, these are used in the computer program for identification purposes. A printout record of inspection results can be obtained at any time (i.e. hourly, per shift, daily or monthly) and from this, Statistical Control information is available as required.

d) While the blind holes are being probed, the camera and vision system looks at the large holes for the presence of flash and makes an accept/reject decision accordingly. Any flash along an edge wider than 1/32", causes the part to be rejected but does not abort the system. Again, this information is displayed on the CRT and can be obtained in the same way as with the trigger probe.

4.3 Equipment Used in the Workcell

The following equipment was used in the ORC automatic inspection workcell:

MERLIN Robot

The Merlin Robot, manufactured by American Robot Inc., is modular in design and configurable up to 6 axes. It has a cast aluminum structure with 6 motor driven axes, and can be mounted on a sturdy steel base, a movable track, or installed on an overhead gantry. Each axis is driven independantly by a precision, closed-loop stepper motor. The repeatability of the MERLIN Robot is $\pm$0.001", with a maximum payload of 20 lb. With this payload the maximum tip velocity at the 60" reach is about 5 feet per second.

Robot Controller

The Robot controller makes use of eight microprocessors, and has an integrated 32 bit/16 bit computer architecture. The system is controlled by a host Motorola 68000 microprocessor with slave Motorola 6809 microprocessors, one for each axis, and one for the disk drive control. The system is supported by 32 discrete digital I/O channels, 2 analog ports and 4 RS232-C serial ports. The resident soft-ware is AR-BASIC and AR-Smart, both of which are easy-to-use programming languages. A hand-held teach pendant, with joystick and a keypad, can also be used to program the Robot.

AR-Basic is a version of the basic language adapted for use by American Robot for programming the Merlin. This system allows development, editing and storage of computer programs. Using AR-Basic is similar to programming a personal computer and gives the Robot a high degree of flex-ibility. AR-Basic must be used in order to print reports and summaries.

Robot End Effector

The robot end effector is shown in figure 3 and comprises a miniature video camera mounted above and in-line with a touch-trigger probe.

Automatic Vision System

The Autovision 4 machine vision system manufactured by Automatix Inc., consists of the AV-4 Controller which can control up to eight video cameras (16 optional). The standard unit contains two central processors and a vision processor board. Additional slots are provided in the con-troller backplane for upgrading the systems' capability. The built-in monitor (CRT), is a solid state unit providing a high quality display of real video and processed video images. The control panel includes a camera selector switch, brightness and contrast controls,

control mode switch and a full typewriter type of keyboard. The Automatix vision system is programmed in the RAIL language which is similar to Pascal. This language is very structured and very powerful, how-ever, it is somewhat more difficult to use than the Basic language.

Hitachi Camera

The Hitachi #KT-230 miniature video camera is a solid-state, black and white, high resolution imaging device. The image pro-duced has 384 (V) X 485 (H) picture elements giving an equivalent resolution of 450 (H) X 359 (V) lines.

Renishaw Touch-Trigger Probe

The probe is an electronic triggering device which provides a signal indicating that the stylus has been deflected in any of the directions: $\pm$X, $\pm$Y or +Z. Once the stylus is deflected in any of these directions by more than 0.0001", an out-put pulse is generated by the probe. It is this pulse that is monitored by the robot controller when the blind holes are being inspected.

Probe Interface

The Touch-Trigger Probe comes with a separate signal processing unit or in-terface (which is mounted on the upper arm of the robot). In addition to pro-viding a normally open contact for inter-facing to the controller, the unit also provides an audible indication of probe actuation. As the probe is deflected, all output signals change state and the aud-ible indicator delivers a short 'bleep' indicating a "blocked" condition.

Inspection Fixture

The function and construction of the inspection fixture has been discussed in section 4.1(g) under Design Considerations. A schematic representation of the inspec-tion fixture is shown in figure 4.

Backlight Table

Correct lighting of the object for in-spection purposes is critical for a vision system. Lighting must bring out the features of the part, providing sharp contrast and detail of the image. Since ambient lighting provides insufficient contrast and detail, backlighting was selected for this application. Accordingly an 'L' shaped light table is used to satisfactorally illuminate the part.

5.0 Experience With the Workcell

5.1 Merlin Robot

The Merlin robot performed very satisfactorily throughout this project and proved to be a pleasure to work with. At present, in North America, only two robot manufacturers are claiming repeatabilities down to 0.001". Apart from American Robot, the other manufacturer is Intellidex. The Intellidex robot was also considered for this task but, when examined, on a number of occasions, the robot motion did not appear to be smooth and a slight degree of jerkiness was always present. The Merlin, on the other hand, proved to be very smooth in motion, a necessary characteristic if, at some future development stage, pictures are to be "grabbed" while the robot and camera are in motion.

For most of the time, the robot was run at about 25% of its full speed when travelling between holes. At this speed, it took just under 7 minutes to probe the inside of the most complex instrument panel. When flipped, the outside of the panel could be probed in under 4 minutes as there are less holes. For production, the "between hole" speed can be doubled, although the speed at which the probe is inserted into a hole is intentionally low and must be maintained that way in order to achieve the necessary control and accuracy. With this type of speed adjustment the above probing times can be reduced to approximately 5.5 minutes and 3 minutes respectively.

5.2 The Fixture

The fixture used in the workcell (see figure 4) was designed and fabricated to answer a number of requirements. It had to provide a quick and simple means of loading a panel onto a fixture, clamping it in place and quickly rotating the panel between two positions so that the front and back of the panel could be inspected. Given how little detail design effort was put in to the initial design, this fixture performed remarkably well. It should also be noted that the vacuum cups, used to position the panel and hold it in place, operated very successfully.

For the production model inspection fixture, however, a more detailed "custom design" will be made for each type of panel to be inspected. Modifications to be included in this design are: an increase in the number of support and locating points and an increase in the number of suction cups strategically located around the panel. Apart from repeatability and accuracy, these modifications will ensure a more rigid location of the panel on the fixture, eliminating the possibility of movement, or deflection of unsupported areas (eg. as in the case of slight warping due to stresses in the panel). Repeatable and rigid positioning of the panel was found to be crucial to this type of inspection workstation, as some variations in position significantly affected the performance of both the probing and machine vision operations.

5.3 The Renishaw Touch-Trigger Probe

This probe performed flawlessly throughout the project. Although very sensitive (0.0001" deflection causes triggering), the probe also proved to be very robust owing to the comparitively large amount of overtravel provided by the probe mechanism (0.2" in the Z direction and 0.5" in the X and Y directions). This type of probe was available from two suppliers in North America, the other being Digital Technics Inc. The probe offered by this company had a novel form of communication channel, in that no wires were needed. Signals were transmitted from the probe by pulsed LED's mounted around the probe. An infra-red receiver, located on the signal processing module, had then to be mounted in a location that was always within line-of-sight of the probe. This latter requirement was seen as a definite disadvantage for this type of operation. This drawback, plus the fact that the Renishaw probe was less than half the cost of the other type, made the decision of which probe to use very easy. The Renishaw probe, manufactured by Renishaw Electric in the U.K., will continue to be used in the production version.

5.4 Machine Vision

Method of Inspecting Openings

Because open holes and their edges were being inspected, backlighting of the panel was used. Thus, when seen as a video image, all the holes are white and the panel appears black. Backlighting is particularly advantageous as it enables a high contrast ratio to be obtained between the hole and the surrounding plastic. Having high contrast greatly reduces the possibility of areas of plastic being interpreted as holes. This kind of confusion occurs without backlighting, because the hole and plastic areas are much closer together on the grey-scale. Highlights, or variations in lighting of the plastic, or shadows behind the hole, can then cause dark shades to become lighter and light shades to darken. In a situation like this, the vision system can not resolve which is hole and which is plastic. Thus, by increasing the contrast, a wide range of the grey-scale can be maintained between the "white" holes and the "black" plastic, and imaging ambiguities can be avoided.

The method of detecting flash was to program the vision system to draw "windows" around the edges of the holes to be inspected. Once these "windows" have been defined by the software, all subsequent vision processing is restricted to the image within these windows and everything else is ignored. Flash was then detected by calculating the ratio of the white area within the window, (hole), to the "windowed" black area, (plastic edge). An increase in the area of black, above a certain tolerance value, indicated the presence of flash.

As will be appreciated, the effectiveness of this method relied heavily upon the positional repeatability of the robot and the part locating repeatability of the fixture. In both cases repeatability errors were sufficiently small that they had a negligible effect on the performance of the visual inspection systems.

5.5 Optical Equipment

A number of camera and lens combinations were tested before making the final selection of optical equipment:

Panasonic Camera:

A Panasonic solid state video camera, using charge coupled device (CCD) technology, is the standard camera supplied by Automatix. This camera weighs about 4 pounds and is approximately 8"x4"x2" in size. Inspection of the holes was originally attempted using this camera. However, the size of the camera proved to be too bulky when mounted at the end of the robot arm and limited the manoeuvering ability of the robot during probing.

Panasonic Camera and Fibrescope

To overcome the problem of bulkiness at the end of the robot arm, the camera was relocated to the upper arm of the robot. A fibrescope was then connected to the camera such that the fibrescope lens was mounted just above the probe. An 11mm diameter Olympus fibrescope was used, having an overall length of 2 metres.

This type of fibre-optic device is normally used for inspection tasks where a high resolution image is required and the access path is either very narrow or curved (eg. boiler tubes).

The use of fibrescopes to provide an "eye-in-the-hand" type of vision system is not a new idea and has been implemented on other robots. However, when tried in this application, it quickly became apparent that the large amount of manoeuvering of the robot arm and wrist was causing considerable flexing, and some twisting of the fibrescope. Because of its high fibre density and fibre-bundle construction, this type of device can not be subjected to twisting without damage occurring. Thus, the use of fibrescopes was not recommended for this application.

Miniature Hitachi Camera

A miniature CCD video camera, manufactured by Hitachi (model No.KT230), was also evaluated for this application. This camera weighs just over one pound and is approximately 3"x1.5"x1.25" in size. The resolution capability of this camera is better than that of the Panasonic camera, giving approximately 20% more lines per frame, both horizontally and vertically. It should be noted, however, that the final resolution of the vision processed image was determined by the Automatix machine vision system's capabilities and, therefore, was identical in all the optical testing.

The big advantage of using the Hitachi camera was that it could be located directly on the end effector of the robot. Because of its reduced size, it could be mounted just above the probe and did not interfere with, or limit, the robots motions during inspection. Also, the camera was small enough that the robot could move the lens right up to a hole (for better resolution) without physically interfering with the panel.

Testing commenced using the standard 25mm lens supplied by Hitachi with the camera. Because of the limited depth-of-field of this lens, it was necessary to operate the camera from a standoff distance of approximately 15". At this distance it was only possible to detect the presence of gross flash, ie. approximately 0.125" wide by 0.5" long. To detect the presence of flash that is at least 0.030" wide would, therefore, require an improvement in resolution, both optically and in the processing of the image.

The camera manufacturer was contacted and various combinations of lens types and spacing rings were evaluated. Of the lenses tested, an AUTO-IRIS lens was found to provide very satisfactory results. Use of the AUTO-IRIS lens enabled much better control of image brightness and greatly improved the depth-of-field. The improved image enabled the vision system to do a much better job of resolving the edge of the part, while the greater depth-of-field allowed the camera to be moved much closer to the holes (typically, a 6" standoff).

Repeated testing, using the AUTO-IRIS lens, enabled flash to be detected down to 0.035" along an edge which was at least 2" in length. Therefore, to determine the presence of flash around the complete edge of a large hole, required the camera to be moved to several points around the hole perimeter. At each point, the camera would then take pictures of 2" segments of the hole's edge. This type of hole segmentation and robot movement obviously adds a burden to the total inspection time of the system. However, this burden is samll enough that machine vision inspection of all the major openings in the panel could still be completed in under 16 minutes.

6.0 Future Developments

Although a working automatic inspection station has been built at ORC, plans are now in process to refine the design and build a production unit. This unit could be installed in the plant by the end of June, 1985. As mentioned earlier, part of this unit has already been designed, eg. the gantry mounted robot (see figure 2), the use of four inspection tables and fixtures (see figure 1) and an improved type of fixture. Other modifications presently being evaluated include: testing

of a wider range of camera lenses and other optical devices and evaluation of newer types of hardware based machine vision systems.

These newer vision systems are hardwired, and make use of pattern recognition techniques rather than the standard SRI algorithms used in most software based vision systems (such as the Automatix AV4). The big advantage of this type of vision system is speed, being orders of magnitude faster than the software systems. Also, because this type of system makes use of pattern recognition, it may be possible to inspect the openings without segmenting the hole into so many parts, thus speeding up inspection time. It is thought that by optimizing various operations of the robot and machine vision system, the present total inspection time of 27 minutes can be reduced to under 20 minutes for the production unit.

7.0 Conclusion

A workcell has been designed, built and tested at ORC to perform automatic inspection of plastic moulded automobile instrument panels. The part used in the test had 60 blind holes and approximately 15 major openings that had to be inspected. The blind holes were checked automatically for presence and adequate depth, while openings were automatically inspected for flash around edges. The method of testing successfully incorporated the use of tactile sensing and machine vision into the end effector of a robot. Tests have shown that the inspection time for a panel can be halved and the need for an inspector eliminated by use of this type of inspection workcell. Design is now proceeding on a production unit for installation in 1985.

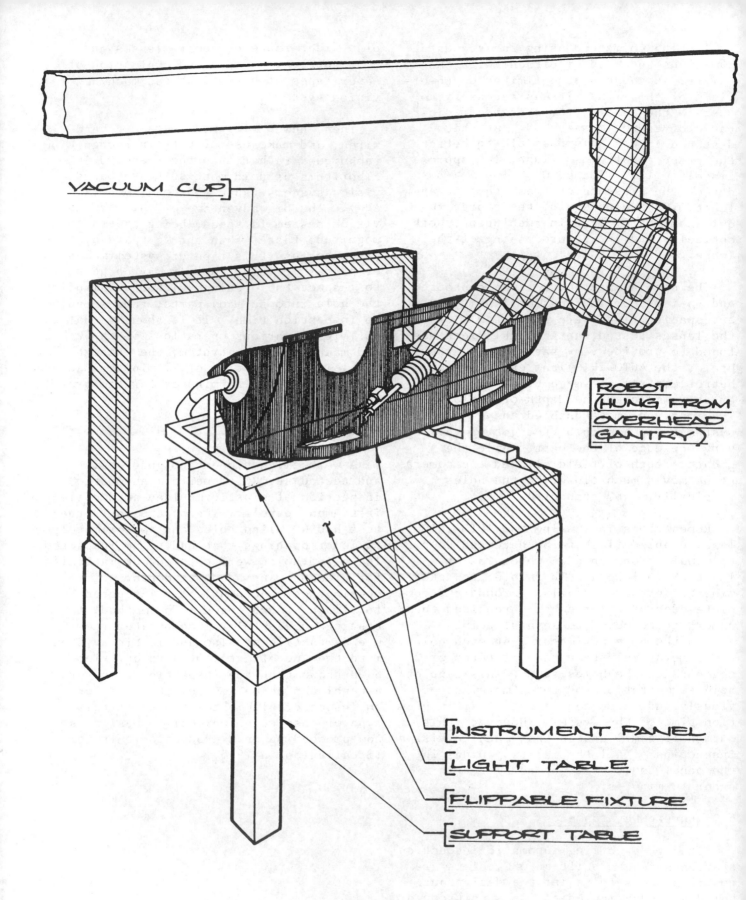

VACUUM CUP

ROBOT
(HUNG FROM
OVERHEAD
GANTRY)

INSTRUMENT PANEL

LIGHT TABLE

FLIPPABLE FIXTURE

SUPPORT TABLE

PROPOSED ARRANGEMENT OF
ROBOT, PANEL FIXTURE AND
BACKLIGHT SUPPORT TABLE

FIG. 2

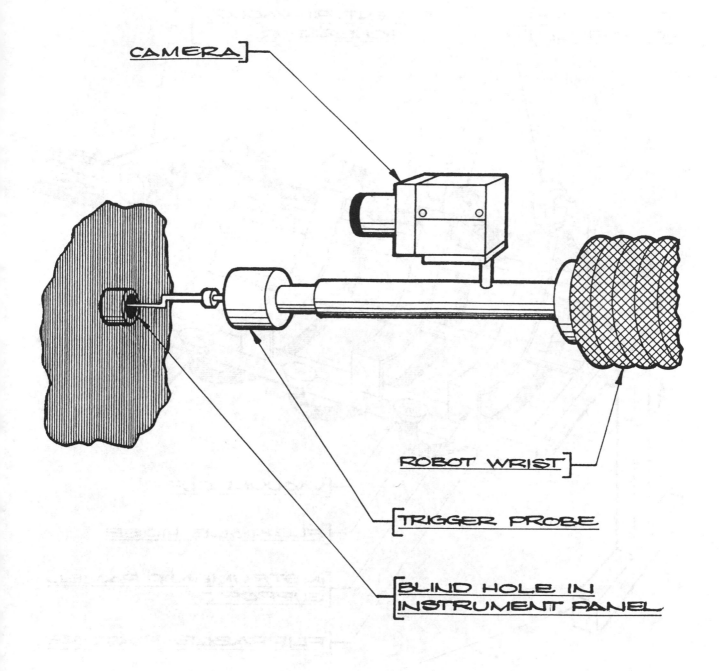

CAMERA

ROBOT WRIST

TRIGGER PROBE

BLIND HOLE IN
INSTRUMENT PANEL

PROPOSED ARRANGEMENT OF
TOUCH-TRIGGER PROBE & VIDEO
CAMERA AS END EFFECTOR OF ROBOT

FIG. 3

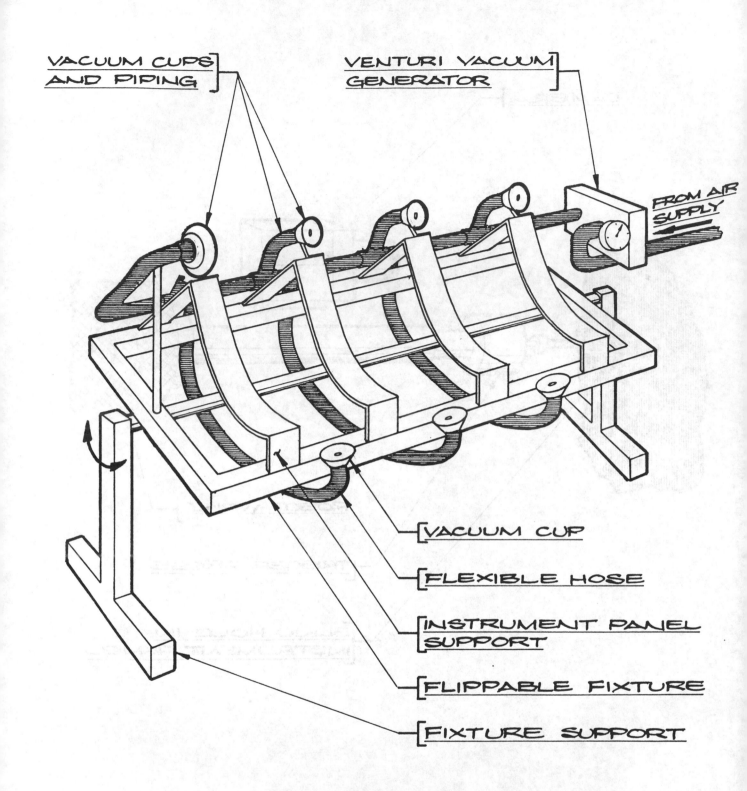

VACUUM CUPS AND PIPING

VENTURI VACUUM GENERATOR

FROM AIR SUPPLY

VACUUM CUP

FLEXIBLE HOSE

INSTRUMENT PANEL SUPPORT

FLIPPABLE FIXTURE

FIXTURE SUPPORT

BASIC FRAME OF TWO-POSITION FLIPPABLE FIXTURE FOR HOLDING PLASTIC INSTRUMENT PANELS

FIG. 4

258

THE IN-PROCESS QUALITY CONTROL BY NEW TECHNOLOGY, CAQC

H. Yamada, Mitutoyo Asia Pacific Pte. Ltd. Singapore

The most advanced technology, CAD/CAM aims at the objective of improving productivity since it is a common object for every manufacturer to produce goods of high quality at a low cost.

With this situation, the innovation of quality-control technique is required to keep pace with more sophisticated manufacturing technology.

The solution is SPC (Statistic Process Control) or CAQC (Computer Aided Q.C.) developed by MITUTOYO.

1) What is SPC (The concept of SPC)?
2) The reason why SPC is being introduced.
3) SPC system
    a) Stand alone system
    b) Batch data processing system
    c) On-line data processing system
4) Future prospect for SPC (the important relations between CAD/CAM system and CAQC system).

The most advanced technology, CAD/CAM aims at the objective of improving productivity since it is a common object for every manufacturer to produce products of high quality at a low cost.
Under this situation, the innovation of high quality-control technique is required to keep pace with more sophisticated manufacturing technology.
The solution is SPC or CAQC developed by MITUTOYO.

Every manufacturer commonly endeavours to produce its products with the highest possible quality and at the lowest possible price to serve best its clients, which can successfully be performed largely by enhancing the productivity. It is usually said possible through standardization and stabilization of so-called 4M; Material, Machine, Man and Method.

A Quality Control Division plays an important role for this purpose. If a conventional method of quality control is followed, too much time will be wasted while transferring measuring data statistically which again is fed back to workshop, to catch up with the high speed at which a volume of components is turned out by modern automated machines such as NC machines, CNC machines or machining centers.

A failure in this proper control may bring forward a number of defective products if the machining conditions should go meanwhile beyond the control limit. In order to prevent such a waste, the importance increases in recent years to materialize "instant quality control on the spot" by a machine operator in a production line. Through review of a whole quality-control system, a new concept of "Quality is built in process" is being established.

WHAT IS SPC OR SQC?
SPC is a shortened form of Statistical Process Control, meaning the same as SQC (Statistical Quality Control), which was advocated by DR. W. Edwards Deming in 1920s in U.S.A.

The point of his assertion may be interpreted in the following way; "When producing products of a high marketability in an economical manner, the theory of Statistical Quality Control proves tremendously useful. Every possible principle and method of statistics should be effectively applied to all steps of production." This in other words means "Production processes can be controlled by statistically processed data of measurements while the quality of products simultaneously controlled in the production processes".

After the World War II in Japan, this SQC concept has been widely adopted with successful results in a large number of industries. This however needs to be further improved to materialize a real-time-base control system especially in order to shorten the time to the possible minimun in the process of production → measurement → data process → information feed back → machine correction. (For the purpose of immediate evelution, interpretation and appropriate process action).

The new method may be considered as a way to Defect Prevention itself as against the traditional idea of Defect Prevention Control System.

The disadvantage of defect detection is that out- of-tolerance components must be produced before the operator can determine how to adjust the process.
The defect prevention minimizes the possiblilty of producing out-of-tolerance components.

PRESENTATION OF DATA BY MEANS OF SIMPLE FUNCTIONS OR STATISTICS
Table 1 presents for ready reference a list of those functions or statistics, the ones marked by an asterisk (*) being the most importance in the theory of quality control.

Fraction within Certain Limits	Measures of Central Tendency	Measures of Dispersion	Measures of Lopsidedness or Skewness	Measures of Flatness or Kurtosis	Measures of Relationship or Correlation
* Fraction defective p	* Arithmetic mean $\bar{X}$ $\dfrac{\text{Maximum} + \text{Minimum}}{2}$ Median $\tilde{X}$ Mode	* Standard deviation σ Variance σ^2 Mean deviation Observed range R	* Skewness k	* Flatness B_2	* Correlation coefficient r Correlation ratio η

Table 1. Commonly used functions or Statistics

Arithmetic mean : $\bar{X}$

This measure is commonly used for indicating central tendency and it is obtained from the formula stated below.

$$\bar{X} = \frac{\sum\limits_{i=1}^{n} X_i}{n},$$

where, X_i: Individual measurement data
n: Number of measurements

Standard deviation: σ

The most common measure of dispersion or spread is the standard deviation.

$$\sigma = \sqrt{\frac{\sum\limits_{i=1}^{n} (\bar{X} - X_i)^2}{n-1}} \quad \text{(sample)}$$

Obviously, a small standard deviation usually indicates that the values in the observed set of data are closely clustered about the arithmetic mean, whereas, a large standard deviation indicates that these values are spread out widely about the arithmetic mean.

Fig. 1. shows the two continuous distributions of same functional form, deferring only in the standard deviation.

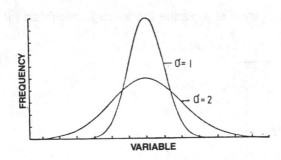

Fig. 1. How the Standard deviation σ indicates dispersion

Skewness: K

The skewness of a distribution of n values of X is designated by the letter K and defined by the expression,

$$k = \frac{\dfrac{\sum\limits_{i=1}^{n} (X_i - \bar{X})^3}{n}}{\left[\dfrac{\sum\limits_{i=1}^{n} (X_i - \bar{X})^2}{n}\right]^{3/2}} = \frac{\dfrac{\sum\limits_{i=1}^{n} X_i^3}{n} - \dfrac{3\bar{X}\sum\limits_{i=1}^{n} X_i^2}{n} + 2\bar{X}^3}{\sigma^3}$$

K may be either positive or negative.
If the distribution is symmetrical, K is zero. Fig. 2 shows two continous distributions of the same functional form, differing only in skewness

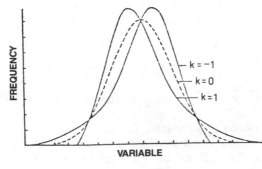

Fig. 2. Illustrating use of k as a measure of skewness

Flatness: β_2

The flatness of the distribution is defined by the expression.

$$\beta_2 = \frac{\dfrac{\sum\limits_{i=1}^{n} (X_i - \bar{X})^4}{n}}{\left(\dfrac{\sum\limits_{i=1}^{n} (X_i - \bar{X})^2}{n}\right)^2} = \frac{\dfrac{\sum\limits_{i=1}^{n} X_i^4}{n} - 4\bar{X}\dfrac{\sum\limits_{i=1}^{n} X_i^3}{n} + 6\bar{X}^2\dfrac{\sum\limits_{i=1}^{n} X_i^2}{n} - 3\bar{X}^4}{\sigma^4}$$

Fig. 3. shows three symmetrical frequency distributions difering only in the degree of flatness.

261

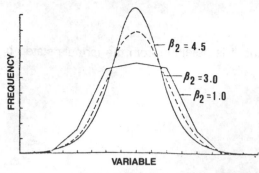

Fig. 3 Illustrating use of β_2 as a measure of flatness of distribution.

We may divide observed distributions into two classes — those that have and those that have not arisen under controlled conditions.

For the distributions of first class, the three simple statistics, average $\overline{X}$, standard deviation σ, and skewness K contains almost all of the information in the original distribution.

For those of the second class, the most useful statistics are the average and standard deviation.

These contain a large part of the total information in the original distribution, at least in respect to the number of observations lying within symmetrical ranges about the average.

COMMON STATISTICAL TOOLS

Histograms
A bar graph displaying a frequency distribution.
As a picture of the quality of the sample, it may be used to show at a glance the average quality and the comparsion of the quality with specific requirements.
This tool is used in the analysis of the quality of a given process or product.

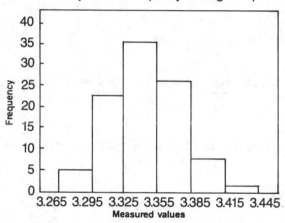

Fig. 4 Typical Histogram

Control Chart

A method of monitoring the output of a process or system through the sample measurement of a selected characteristics and analysis of its performance over time.
This tool may be used to maintain control over a process after the frequency distribution has demonstrated that the process is "in process".

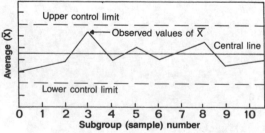

Fig. 5. Typical Control chart, $\overline{X}$-chart

Of the analystical tools, the control chart is a particularly useful and effective technique.
The Shewhart control charts are some of the most significant and widely used tools for process control and they effectively direct attention toward special causes of variation that must be reduced by management action.

Several types of control charts have been developed to analyse both variables and attributes, however, all control charts have the two same basic usages.

1) As a judgement, to give evidence whether a process has been operating in a state of statistical control, and to signal the presence of special causes of variation so that the corrective can be taken.
2) As an operation, to maintain the state of statistical control by operating the control limits as a basis for real-time decisions.

X̄ − R chart

An X̄ and R chart, as a pair, are developed from measurements of a particular characteristic of the process output.
These data are reported in small subgroups of constant size with subgroups taken periodically (eg. once every 15 minutes, twice per shift etc.)
The characteristics to be plotted are the sample average (X̄) and the sample range (R) for each subgroup, these reflect the overall process average and its variability respectively.

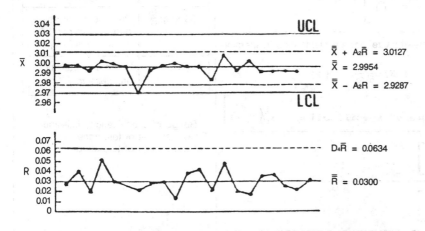

Fig. 6. X̄-R control chart

X̄-S chart

X̄ and S charts are developed from measured process output data and are always used as a pair.
The sample standard deviation S is a somewhat more efficient indicator of process variability, especially with larger sample.

CONTROL CHARTS FOR ATTRIBUTES

Attributes-types data have only two values (confirming/unconfirming, pass/fail, go/no-go, present/absent) but they can be counted for recording and analysis.
The next four subsections cover the fundamentals of the four major types of attributes control chart.

P chart indicates the proportion of units non-confirming (from samples not necessarily of a constant size)

Table 2. **COMPUTATION METHOD OF PROCESS CAPABILITY**

KIND	ITEM	DESCRIPTION	DATA USED
CONTINUOUS DATA	PROCESS CAPABILITY	Quality distribution described in amount Range of dispersion of $6\sigma p$ (mean value X̄p)	Data subjected to variation by 4M 1. Histogram 2. Histogram of each value on X̄ control chart (on determination of σp from estimation on R) 3. Process Capability Chart → Histogram
	MACHINE CAPABILITY	Quality distribution attributable to machine Range of dispersion $6\sigma m$ (mean value X̄m)	Data eliminating variation except machine factor 1. Continuous control chart $$\sigma m = \frac{\bar{R}}{d_2}$$ 2. Process capability chart → Histogram 3. When a trended variation notices on the above, $$\sigma m = \frac{\bar{R}S}{d_2}$$ R̄s: average moving range
ENUMERATED DATA	PROCESS CAPABILITY	Average fraction defective P Average number of defects c Number of defects per unit u Average number of defective units units pn	Control chart usually for a period over a month P control chart C control chart U control chart Pn control chart

Table 3. COMPUTATION AND EVALUATION OF EXPONENT OF PROCESS CAPABILITY

	COMPUTATION FORMULA	APPRAISAL
Range of Tolerance	1. Dispersion alone is a problem whereas mean value is easily adjusted $$C_p = \frac{T}{6\sigma_p}$$ $$\left[C_m = \frac{T}{8\sigma_m} \quad (1) \right]$$ 2. Difficult in adjusting mean value, and problems lie in dispersion and mean value $$C_{pk} = (1-K)\frac{T}{6\sigma_p} \quad (2)$$ $$\left[\begin{array}{l} K \geqq 1, \; C_{pk} = 0 \\ K = \left\| \text{middle value of specification} - \bar{x} \right\| / (T/2) \end{array} \right]$$	<table><tr><td>Cp – value</td><td>Judgement</td></tr><tr><td>Cp > 1.33</td><td>Process capability satisfactory</td></tr><tr><td>1.33 ≧ Cp ≧ 1</td><td>Process capability normal, but careful management required</td></tr><tr><td>1 ≧ Cp</td><td>Process capability unsatisfactory</td></tr></table> T: Range of specification, Tolerance Tu: Upper limit of tolerance TL: Lower limit of tolerance
Upper Limit	$$\left[C_p = \frac{Tu - \bar{x}}{3\sigma_p} \right]$$	
Lower Limit	$$\left[C_p = \frac{\bar{x} - TL}{3\sigma_p} \right]$$	
Remarks	(1) $c_m = \dfrac{T}{8\sigma_m}$ Calledas Machine capability index for convenience (2) In the case of Cpk 1, when Cpk value alone is insufficient to judge if it is bias or dispersion of mean value, it is better to be combined with Cp.	

Np-chart indicates the number of units non-confirmities (from sample of constant size).

C-chart indicates the number of non-confirmities (from sample of constant size).

U-chart indicates the number of non-confirmities per unit (from sample not necessarily of constant size).

MEASURING TOOLS AND STATISTICAL ANALYNZER FOR SPC

It has been explained so far how it is advantageous to adopt the SPC theory based on its thorough understanding. Fo exercising it practically however some difficulties lie in shortening the time needed to process the data statistically involving the production line without delay. These difficulties can be solved easily by M-SPC (Mitutoyo Statistical Process Control System)

The main aim of this M-SPC is to eliminate entirely any possibility of non confirmity articles in the production line itself which certainly will increase productivity a great deal. This enables a production operator to perform measurement, automatic judgement of GO — N.G. and statistic data process, whose results will be described in the forms of Histogram, $\bar{X}$-R char or $\bar{X}$-S chart to control the production accuracy to the optimum level.

M-SPC is available in three different systems; Stand Alone System to process data with directly input measuring data, Batch Data Processing System to centralize process and control by a host computer connected with plural numbers of data, and On-line Data Processing System. These can be choosed to suit the production system of each company.

M-SPC System basically comprises of measuring instruments with data output function (Digimatic Caliper, Digimatic Micrometer Digimatic Indicator, etc.) and Micro-processor based Statistical Analyzer or Host Computer to print out the statistically processed data.

STAND ALONE SYSTEM

This system can be used for quality control and production control right in the production line, associated Mini-digimatic Processor DP-1 or DP-2 as the statistical analyzer.

DP-1

The measuring data can be directly printed out in the production line to show mean value ($\bar{X}$), maximum value (Xmax) minimum value (Xmin), range (R) and standard deviation (σ) computed against a given number of measurements (N).

This system indicates out-of-tolerance results by marking Δ or ∇ when preset to tolerancing. A histogram is also automatically printed out with the tolerance range divided to 10 division. The maximum data input points is 1000.

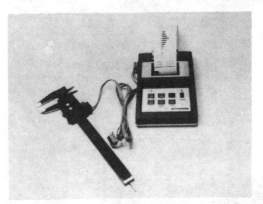

Fig. 7. Statistical Analyzer, DP-1 with measuring tools (digimatic Caliper and digimatic Micrometer)

DP-2

Besides various kinds of data processes required to prepare $\overline{X}$-R Control Chart, this system classifies five different grades of acceptance-or-rejection results, giving on alarm signal in the case when the limit 2 repeats three times continuously. This make possible an instant control of the condition of production machines or sorting equipments. In addition, compensation of thermal effect on measurement data will be exercised against measured value for statistical data process. This also can compute a measured value by a constant or by the four fundamental rules of arithmetics, which can instantly process any data eliminating human computation errors. Process Capability Index can at the same time be computed and printed out.As in DP-1, the maximum input point is 1000 data.

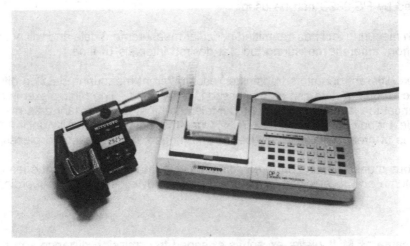

Fig. 8. Statistical Analyzer, DP-2

Additionally to those, there are available further advanced system such as DP-3 and DP-20 to process more complex statistical data as statistcal analyzer. Three different measuring instruments can be connected at the same time to either DP-3 or DP-20. DP-3 can compute out estimate defective ratio (ZUSL, ZLSL) and print out $\overline{X}$-R chart and $\overline{X}$-S chart.

Furthermore, DP-20 can work out skewness or flatness, and programming by a user is possible. Data output can be connected to a floppy disk, a printer, a plotter or a CRT.

BATCH DATA PROCESSING SYSTEM

If the batch data peocessing is required by a host computer for centralized process control, the measurement data can be stored by the data logger which can be carried to the location of measuring tools in a workshop. The data thus stored in a data logger can be transmitted through a transmitter to the host computer interfaced by RS-232C.

Maximum logging number: 1000
Settable characters: 1 — 10

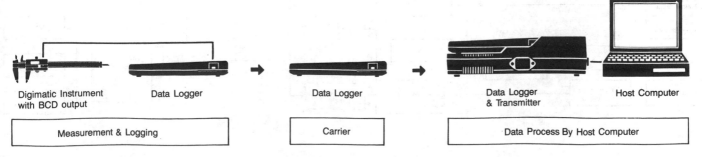

| Digimatic Instrument with BCD output | Data Logger | | Data Logger | | Data Logger & Transmitter | Host Computer |

| Measurement & Logging | Carrier | Data Process By Host Computer |

Fig. 9. Batch Data Processing System

Fig. 10. Data Logger

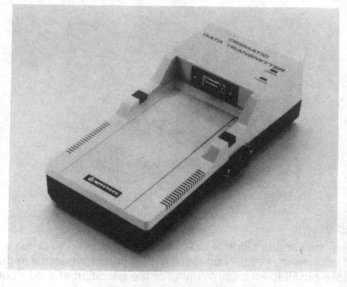

Fig. 11. Data Transmitter

ON-LINE DATA PROCESSING SYSTEM

If the direct transmission is preferred to from the measuring tools to the host computer for on-line data processing, three channel Multi-plexer interfaced by RS-232C can be used.

The Multi-plexer can tell which mesuring tool has tranmitted paricular measurement data and will not be confused by various measuring data transmitted from different measuring tools at desired intervals of time.

For instance, even when three different operators transmitted each different measuring data on different measuring items through differnet sources, each required data can be processed independently. The Multi-plexer is provided on the display panel with such functions cancel the last-input data or preset tolerance that unnecessary data may be easily removed or non-conformity articles may be sorted out in a workshop. Thus, when the Multi-plexer is fully untilized, it is no longer a mere interface to gather data, but is of great help for classification and judgement of measured pieces.

Since it can make simultaneous measurements on three different items, the application of the feature can be for instance to measure the area, the volume and the weight, or to obtain measured values fron two-coordinate or three-coordinate measurements.

DATA LINK

The Data Link is almost the same as Multi-plexer except its extended transmission distance and its increased number of connections with measuring tools.

By extending the connecting cable, a measuring tool can be connected with the Host-computer located maximum 1 Km away. It is wonderous to imagin that a manager is looking at control data right on his desk transmitted secondly second from his workshop.

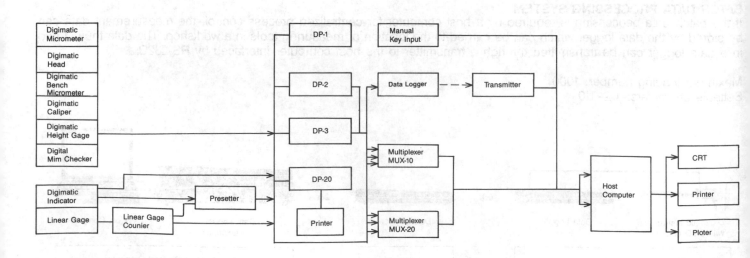

Fig. 12. Mitutoyo SPC System Chart

266

Various kinds of control charts obtained by a Host — computer are shown hereinafter.

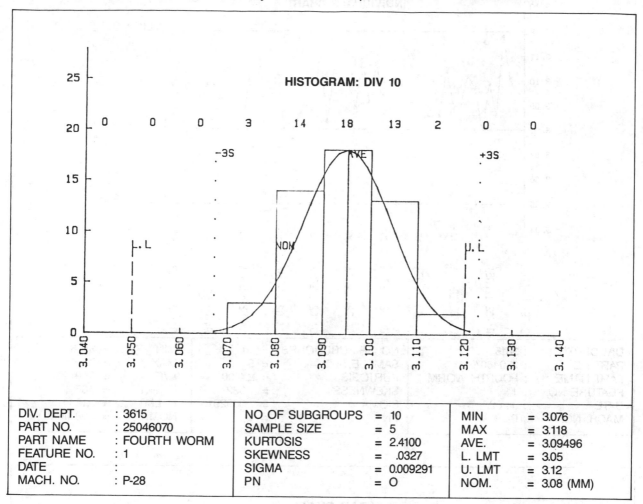

DIV. DEPT.	: 3615	NO OF SUBGROUPS	= 10	MIN	= 3.076
PART NO.	: 25046070	SAMPLE SIZE	= 5	MAX	= 3.118
PART NAME	: FOURTH WORM	KURTOSIS	= 2.4100	AVE.	= 3.09496
FEATURE NO.	: 1	SKEWNESS	= .0327	L. LMT	= 3.05
DATE	:	SIGMA	= 0.009291	U. LMT	= 3.12
MACH. NO.	: P-28	PN	= O	NOM.	= 3.08 (MM)

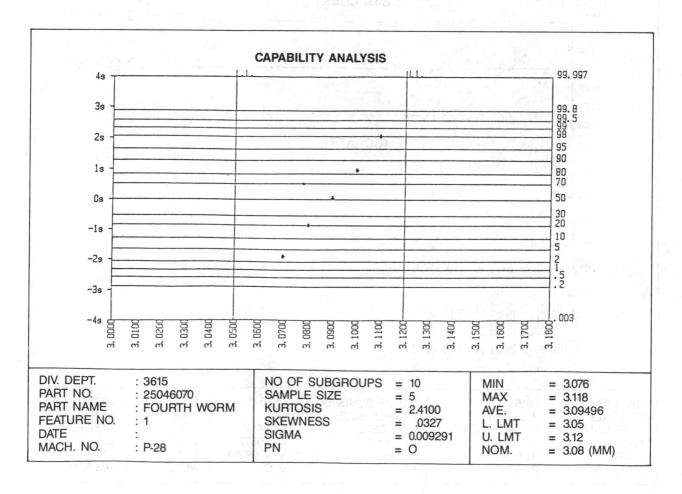

DIV. DEPT.	: 3615	NO OF SUBGROUPS	= 10	MIN	= 3.076
PART NO.	: 25046070	SAMPLE SIZE	= 5	MAX	= 3.118
PART NAME	: FOURTH WORM	KURTOSIS	= 2.4100	AVE.	= 3.09496
FEATURE NO.	: 1	SKEWNESS	= .0327	L. LMT	= 3.05
DATE	:	SIGMA	= 0.009291	U. LMT	= 3.12
MACH. NO.	: P-28	PN	= O	NOM.	= 3.08 (MM)

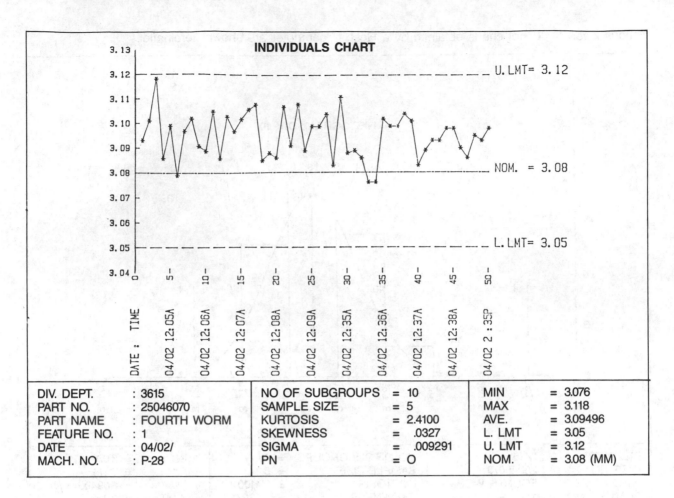

DIV. DEPT.	: 3615	NO OF SUBGROUPS	= 10	MIN	= 3.076
PART NO.	: 25046070	SAMPLE SIZE	= 5	MAX	= 3.118
PART NAME	: FOURTH WORM	KURTOSIS	= 2.4100	AVE.	= 3.09496
FEATURE NO.	: 1	SKEWNESS	= .0327	L. LMT	= 3.05
DATE	: 04/02/	SIGMA	= .009291	U. LMT	= 3.12
MACH. NO.	: P-28	PN	= O	NOM.	= 3.08 (MM)

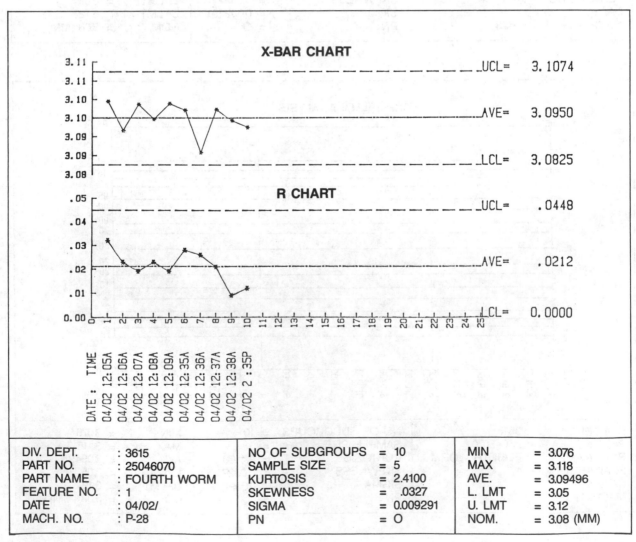

DIV. DEPT.	: 3615	NO OF SUBGROUPS	= 10	MIN	= 3.076
PART NO.	: 25046070	SAMPLE SIZE	= 5	MAX	= 3.118
PART NAME	: FOURTH WORM	KURTOSIS	= 2.4100	AVE.	= 3.09496
FEATURE NO.	: 1	SKEWNESS	= .0327	L. LMT	= 3.05
DATE	: 04/02/	SIGMA	= 0.009291	U. LMT	= 3.12
MACH. NO.	: P-28	PN	= O	NOM.	= 3.08 (MM)

CAD/CAM already is part of essential issue on industrial circle invested a great deal of manpower and cost on software development for computerised CAD/CAM in order to cope with the rapid progress of the industries. It should however be noted seriously that the most ideal checking method of produced 'components' is furnished even when they are produced advanced fully-automated unmanned machining line by CAM. CAQC (Computer Aided Quality Control) therefore should always be associated with CAD/CAM.

An introduction has been made as above to CAQC manually controlled by an operator. The fundamental aim of CAQC however is not only to check dimensions and process data statistically, but to feed processed data back to a production machine to compensate machining conditions automatically without intervention of a man and without breaking the machining cycle.

It will not be long in future when a complete production system is at work in combination of CAM and CAQC.

References
Report — American Machinist special report 762, Januaury 1984
Book — Shewhart Economic Control of Quality of Manufacture Product. 1980
Book — Society of Automative Engineers, Inc. June 1983
Book — Ford motor company, Continuing Process Control and process capability improvement July 1983
Book — Japanese standard association, handbook of quality control. April 1979

MODULAR MECHANICAL ENGINEERING - A REVOLUTION IN ENGINEERING INDUSTRY

Rainer Muck, Jacob AG Mammern, Switzerland

Modular mechanical engineering means a revolution in twice a way:

1. Concerning the classical engineering countries, it initiates a complete new orientation in the engineering industries.

   - from large series to single-piece production
   - from high weights and heavy drive power to intelligent solutions
   - from large working units to small, dynamic workshops

2. Because of the modular construction ot this new machine generation, the mechanical engineers in the consumer countries are now able to construct machines , fitting sophisticated claims. They could not approach this standard in former times.

This opens to countries with a lack of foreign exchange the possibility to restrict very quickly their import of those elements, they are not able to produce themselves.

Almost all machines imitate the movement of a human arm, which directs a tool towards its workpiece, holds it or moves in face of the workpiece. Until some years ago, the imitation of those movements was made linearly. Since the development of roboters circular and combinated rotary movements are possible.(Fig. 1)

Practice shows that a machine and especially its manipulation parts are more rigid with linear movements than with rotary or articulated ones.

However rigidity is the first measure for the accuracy of a machine: If it is not rigid, a triggered point can't be hit exactly; if it is hit, the tool or work piece won't remain in the desired position because of the elasticity of the manipulator.(Fig. 2) This is valid for cutting and non-cutting machines.(Fig. 4B) Also in this last mentioned scope (punching, photographs, welding, soldering, mounting) a higher and higher positioning accuracy is demanded.

The same tendency is evident in the manipulation of tools or work-pieces by roboters; there are demanded higher and higher absolute positioning accuracy and repetitive accuracy as well. This is one reason why the sales of non-linear roboters are not as good as forecast some years ago.

Linear movements can result from
different kinds of driving (screws,
cable pull, cylinder etc.) The
movement will only be linear, if the
geometry of the driving movement has
been transformed into a linear movement; this is done by linear guiding
elements.

The first at the same time rigid
and accurately adjustable linear
guiding element of the industrial
mechanical area is the so-called
dovetail guide.(Fig. 3) For more
than one century it supplied all
qualities necessary for the
production of generations of
industrial goods and commodities.

Much later with the construction of
electric motors, various cylindrical guiding elements were invented.

In the last 20 years the claims for
more accuracy in the fields of
electronics, micro-electronics,
medical and measurement techniques
have increased immediately.
(Fig. 4A)

Accuracy requires high rigidity.If
we talk about rigidity, implicitely
the term "without play" is envolved.

In order to set a sliding, e.g. a
dovetail without play, it is necessary to install a preloading
system, which presses one part of
the guiding element to the other.
If you vary rigidity and material
combination (cast iron/cast iron,
steel/cast iron, steel/synthetic
material), the coefficient of
friction, which determinates the
axial moving force of one guiding
part to the other, is different.

High frictional forces need solid,
strong and therefore expensive
constructions, high axial forces,
and require overdimensioned transmission elements and good greasing.
(Fig.5)
Good greasing is particularly
important for the eduction of heat,
caused by the gliding.

Besides the demand of higher
accuracy, more and more economical
aspects play a part in seeking
solutions for other guiding
systems. Higher technical standard
of the fifties and sixties
permitted the invention and
production of a great variety of
linear guiding elements.

The difference between sliding
guides and rolling guides is a very
low friction coefficient and a very
high rigidity without the loss of
efficiency. For rolling guides
often greasing is not necessary.

The last criterion shows the
possibility to construct much
smaller drive units and correspondingly save power and thermal loss.
Only with the development of
rolling guides for linear movements
it was possible to obtain the

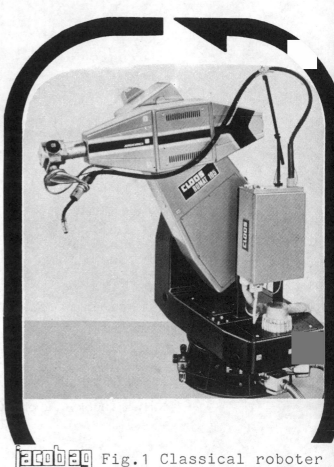

jacob ag Fig.1 Classical roboter

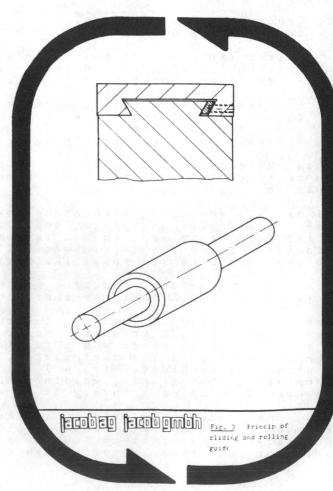

jacob ag jacob gmbh Fig. 3 Princip of
gliding and rolling
guide

272

accuracy as demanded in high technology of mechanical engineering between 1965 and 1978.

The development in the fields of synthetic materials affected guiding techniques as well; nowadays dovetail-slides as well as rails in roll bearing technique can be plastic-laminated.(Fig.6) The result is a guiding element with new friction characteristics, set between a sliding system and a pure metallic gliding element.

This new guiding element distinguishes itself by an outstanding ability of damping. At the same time you don't need rolling-cages for vertical application any more.

Since some years new on the market and applicated more and more are air bearings.(Fig.7) Here the gliding or rolling movement is replaced by an air borne movement. It is frictionless and can be made as rigid as a sliding or rolling guide system.

While in the industrialized countries until some years ago, large series of identical machines were produced (thousands of milling machines, e.g. Bridgeport), this market decreases in favour of a machine market orientated to optimize the application.

Optimization of application means: There must be found a way to construct smaller series and prototypes with similar prices as series machines and the guarantee of short construction and delivery time.

Logical conclusion: These machines are built of parts, which are manufactured in large series themselves.(Fig.8) But, there has to be a wide universality of these elements, and following, the least possible number of purpose-made parts must be produced.

The analysis of always reappearing movements in different machines on one hand, their dimensional and functional analysis on the other hand led to the development of stan-

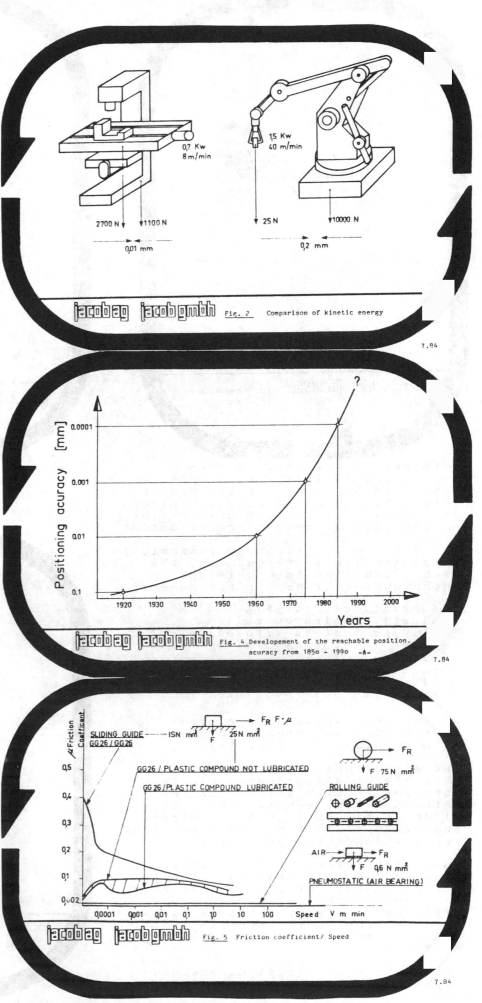

jacob ag jacob gmbh Fig. 2 Comparison of kinetic energy

7.84

jacob ag jacob gmbh Fig. 4 Developement of the reachable position. acuracy from 1850 - 1990 -A-

7.84

jacob ag jacob gmbh Fig. 5 Friction coefficient/ Speed

7.84

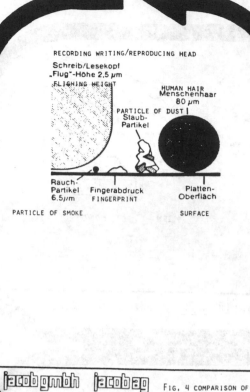

RECORDING WRITING/REPRODUCING HEAD
Schreib/Lesekopf
"Flug"-Höhe 2,5 μm
FLIGHING HEIGHT

HUMAN HAIR
Menschenhaar
80 μm

PARTICLE OF DUST
Staub-
Partikel

Rauch-
Partikel
6.5 μm

Fingerabdruck
FINGERPRINT

Platten-
Oberfläch

PARTICLE OF SMOKE   SURFACE

jacob gmbh jacob ag FIG. 4 COMPARISON OF SIZES
-B-

7.84

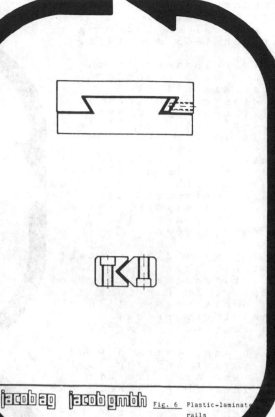

jacob ag jacob gmbh Fig. 6 Plastic-laminate rails

7.84

dardized linear drive assemblies.(Fig.9) They are fit and completed by other, on the market obtainable standardized elements: and allow a mechanical engineering on the base of modular construction.

Based on application and economical specification it turned out, that such assemblies in 4 guiding techniques will do for the moment:

- dovetail slides (Fig. 10)
- plastic laminated rails
- sliding systems with different cages (e.g. cross roller rails, slides)
- sliding systems with a recirculation guide system (e.g. ball shoes and ball bushings)

Now a pneumostatically guided system with an integrated drive has been developped. (Fig. 11)

The dimensional analysis shows, that 85% of all produced machines can operate with slide breathes between 30 and 400 mm. Therefore slide cross sections are designed and produced in the same dimensions for all 4 guiding principals.

Generally strokes are directly dependent on the desired rigidity and accuracy. For a lot of lengths of strokes, that means the difference between the long fixed part and the short wandering part, can be standardized for each cross section.

Equal cross sectional dimensions allow a purchase of material in

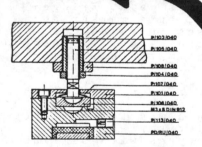

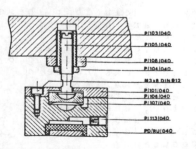

jacob ag Fig. 7 Princip of air-bearings

7.84

great quantities and the production of raw profiles in wide length; this leads to an interesting price for such standardized slides. (Fig. 12A, Fig. 12B) The drive of standardized slides can be made in various ways and is mostly determinated by rigidity and desired positioning accuracy: hydraulic, pneumatic, electrical, mechanical.

Mechanical drive can be made with conventional screws in bronce nuts, ground screws in plastic nuts, ball screws with simple or double nuts and satellite roll screws; but also toothed belt, toothed rack, driving wheel etc. are possible and proper for certain tasks.

In the following we should like to point out a problem of driving techniques of modern manipulators with linear movements and of roboters: failure probability. (Fig. 13A) Parallel to the linear slideway and based on the method of calculation of rotary rolling bearings, ball screws were developped.(Fig. 13B)

The calculation of lifetime of those ball screws is based on the formula

$$L = \left(\frac{C}{F}\right)^3 \times 10^6$$

C = dynamic carrying figures
F = working load

That means a lifetime of 10^6 rotations, when

$\dfrac{C}{F} = 1$

If you calculate the drive of a machine working at high speed, like bonders or high power milling machines, you will recognize, that it is necessary to have huge dimensions as to obtain a reasonable lifetime for ballscrews, respectively to obtain a favourable factor $\dfrac{C}{F}$

plus lifetime.

Geometrically seen this is possible for a high

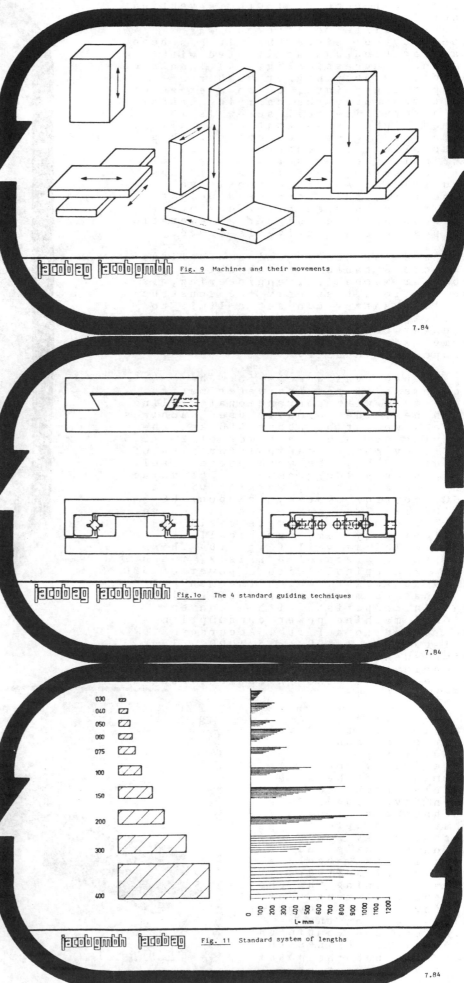

jacobag jacobgmbh Fig. 9 Machines and their movements

7.84

jacobag jacobgmbh Fig. 1o The 4 standard guiding techniques

7.84

jacobgmbh jacobag Fig. 11 Standard system of lengths

7.84

275

power milling machine, but not for a
high power bonder. In the seventies
therefore a variant to a ball screw,
the satellite roll screw, was
constructed; since that it has been
manufactured and applicated with
immense success. (Fig. 14) Thanks of
a considerably higher number of
supporting points, on an average of
13 times more, the carrying capacity
of a satellite roll screw is 13
times higher than that of a ball
screw of the same dimension. This
also allows to construct in an
optimal size the surrounding field
of the screw, that means guidings
and slides. A further positive
aspect is, that less dead weight
has to be carried along during
processing time, higher acceleration
is possible and drive is saved.

The modular construction of slides
and cross tables, which allows a
modular mechanical engineering, is
so manifold, that only the construc-
tor's phantasy can set a limit to
it.(Fig. 15) To illustrate this
modular construction, we will give
some examples out of the different
scopes of industry:

Up to now oval wheels for the
equipment of gas and water meters
were made with conventional milling
machines. The normally used machines
have a weight of about 450 kg, the
driving screw has a power of 2 - 2,5
KW; they move a manipulator mass of
about 110 kg; the work piece itself
weighs only 180 grammes. The value
of such a machine, working with a
CNC- control system, is about US $
46'000.--. The same process however
can be done on a simple small positioning system, numerically
controlled as well.(Fig. 16) This
system consists of 3 axis x, y and z
and of a spindle with a power of 530
W. Dead weight, which the manipulator has to move, decreased to 12,5
kg. In comparison with a conventional machine power consumption
decreased to a fifth. Accuracy
raised from 1/100 to 5/1000, clock
time sunk from 90 to 52 sec; seen as
a whole this means an immense output
increase. Now the investment sunk to
US $ 15'000.--.

It is obvious, that
parallel to the develop-
ment of mechanical
modular systems,
numerically controlled
commands of the same
modular technique have
been developped. This
technique can go so far,
that the simplest type of
a control system consists
of only 1 axis and
contains the absolute
minimum of software.(Fig.
17) Concerning the choice
of the equipment there is
an investment factor of
1:2,5 between the
simplest and completest
execution. Besides the
investor has the possi-
bility to reset or

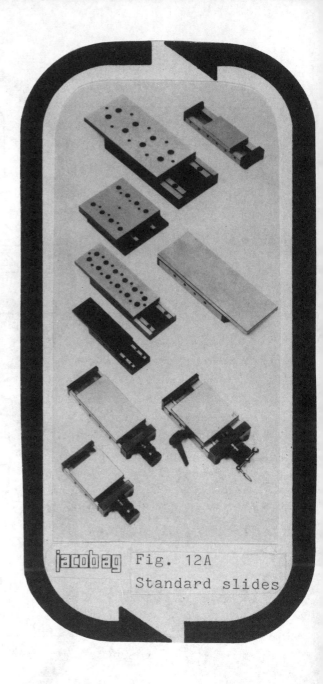

Fig. 12A

Standard slides

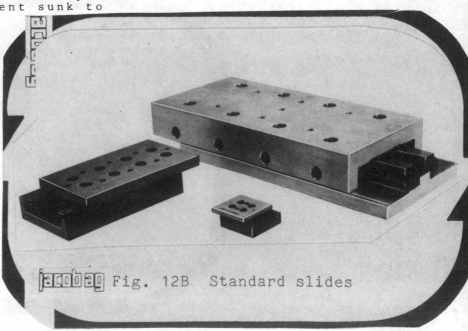

Fig. 12B   Standard slides

276

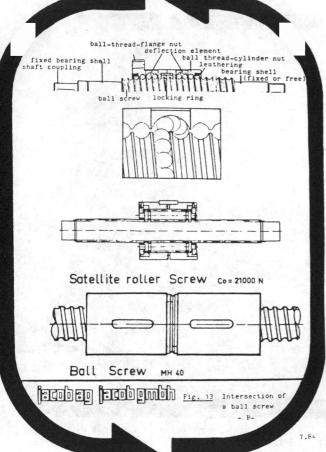

ball-thread-flange nut
deflection element
fixed bearing shell   ball thread-cylinder nut
shaft coupling   leathering
                                  bearing shell
                                  (fixed or free)

ball screw   locking ring

Satellite roller Screw Co = 21000 N

Ball Screw MH 40

jacob ag jacob gmbh Fig. 13 Intersection of
                                  a ball screw
                                  - B -

7.84

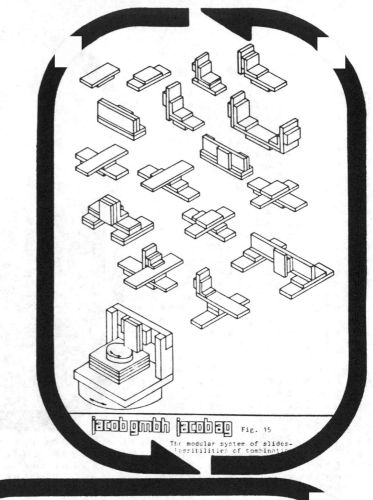

jacob gmbh jacob ag Fig. 15
The modular system of slides-
possibilities of combination

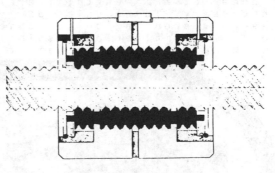

jacob ag jacob gmbh Fig. 14 Sketch of a satallite roller screw

7.84

retrofit the control system for a new product at any time.

The control system CNC 3301 has another advantage: simple video programming and programming after DIN 66025 and 55003. On the other hand it consists of standard cards, because it is constructed in modular technique, and this keeps a first cost installation reasonably low. Example: a 2-axis control system with circular interpolation and a power element for motors of 2 x 5 A. Price: US $ 5'400.--.

Even up to now keys for domestic and motorcar industry are manufactured on conventional CNC- or pantographic controlled machines. Here as well low single weights are guided with the help of an enormous deadweight by a weak spindle (750 W), what is a technical nonsense.

The application of modern engineering techniques, a conventional know how in the techniques of modular precision manipulation units, allow a quick production of fully automatic machines.(8 weeks)(Fig. 18) The value of such a machine consists of more than 50% out of standard positioning systems of various combinations. The machine shown here, works on 3 keys at the same time, with different round heads, but the same opening geometry from raw material to a set of keys in a plastic cup, ready for packaging. Seen from the mechanical side, the investment for this automatic machine with a clock time of 12 sec is not larger than that for 1,5 conventional machines.

A standard numeric command, equipped with special software serves for coding the keys. While in conventional technique this control system is necessary for every single machine, here you need it just once - another advantage of modular construction.

277

A further aspect of
modular construction
becomes more and more
interesting: For the
machine consists from 30
to 90% of modular piece
parts, such machines have
a high retail price and
high reutilization. If
the product, the machine
was planned for, is not
manufactured any more, it
is very simple to
dismount the machine, to
revise the modular as-
semblies, to store the
non-applicable mechanical
special parts for a
possible later resumption
of the same product and
to assemble the modular
parts for a new purpose
to a new machine. More
and more sensitive and
exact parts or parts made
from a special material,
are welded in electron
beam welding chambers.-
(Fig.19)

Fig. 16 CNC-3-axis milling machine
for oval wheels

Because these chambers are in vacuo, you cannot intervene manually, when
the parts have reached the vacuum chamber. All movements must be controlled from outside through windows or cameras. The investment for such
a chamber: US $ 420'000.-- to 1'700'000.--. The parts welded in these
chambers are manifold and range from a round cylindric piece of 6 mm
upto big pieces of 1 m length; therefore several different manipulation
units are necessary. These units are composed of standard slides and
cross tables corresponding to the object, that has to be welded; mostly
you can see 2 - 3 combination possibilities, by which the whole
geometric working area of the chamber can be covered. For these

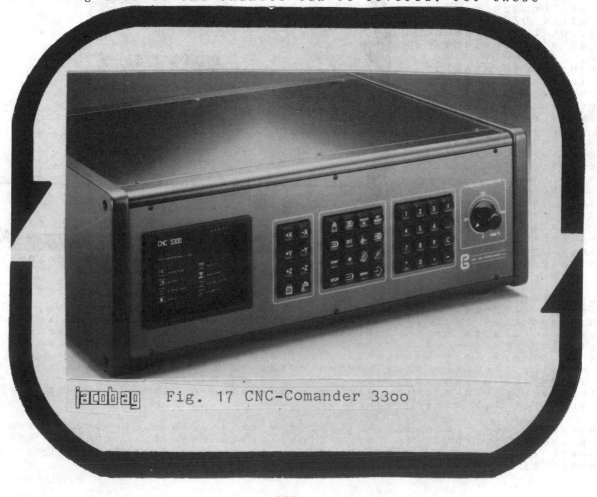

Fig. 17 CNC-Comander 33oo

278

applications in vacuo at 10^{-4} Torr the quality of non-smearing of our guiding elements and screws are the absolute premise.

On the shown installation detonators, with a cross section of 6 mm and an orbit fidelity of 0,02 mm, are palletized (144 pieces) and welded automatically in 0,7 seconds.

The production of contact lenses in skin-friendly, soft synthetic material is only possible, thanks of rigid, extremely accurate guiding techniques.(Fig. 20) Under these given conditions application shows, that such a machine can very easily be mounted on a Diabas plate, serving as amorphous base. The whole mechanism is composed of 2 precision slides, which guide the axial load and trigger the position of the different centres of the lenses, the machine possesses also a rotary movement with a precision roller bearing, prestressed with no play. The dc-motors are charged by the command of all other material processing necessities. This machine works synthetic lenses in a range of 1 micron.

A scratch tester is a testing machine, which is used by film industry to test their products on mechanical resistance.(Fig.22) The machine is composed of a base, 2 sliding guides with screws in x-sense, one special mechanical yoke construction, and one slide in y-sense on the bridge. Here the z-axis serves as tool supporting plate in this portal construction, applicated for most various purposes. In this case the tool is a 5-x pick-up for different material samples, e.g. wood, textile, diamond, gold aso. With these materials the films are scratched; in this way their mechanical sensitivity can be measured and - if necessary - improved.

As technicians and technically interested audience you can easily imagine, that this machine can be equiped with other tools, e.g.

- a drilling spindle to drill printed boards
- cutting spindles for glass fiber industry
- a knife to cut synthetic or glass masks for electronics industry
- a milling spindle matched to the machine for burring industry goods and commodities like e.g. toilet lids

A similar construction has been applicated as a

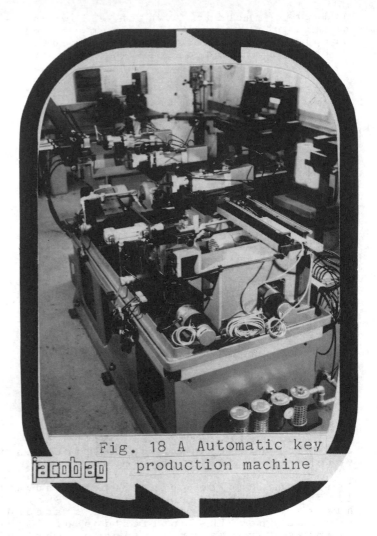

Fig. 18 A Automatic key production machine

Fig. 18B Automatic machine

didactic machine , which allows a
serial working of aluminium, steel
or plastic; it was a great success.
(Fig. 23) It is offered at a very
favourable price to technical
schools, universities, apprentice
schools and factories with apprentice departments. (Fig. 24)

Its modularity and accuracy can also
be applicated in industry, as in
jewellery, where the 12 zodiac signs
and other jewellery pieces in small
series and different materials are
milled and engraved in different
forms.

Here as well a standard control
system can be outfit with engraving
software including the alphabet. It
is a considerable advantage of the
CNC 3300 system, that the dialogue
software used on the terminal can be
programmed quickly and easily in all
the languages of the different
countries. Also it is very simple to
let appear a trade mark in the masks
on the screen, which personifies a
standard command on a serial
applicator.

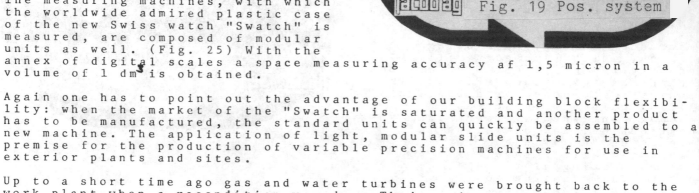

Fig. 19 Pos. system

The measuring machines, with which
the worldwide admired plastic case
of the new Swiss watch "Swatch" is
measured, are composed of modular
units as well. (Fig. 25) With the
annex of digital scales a space measuring accuracy af 1,5 micron in a
volume of 1 dm³ is obtained.

Again one has to point out the advantage of our building block flexibility: when the market of the "Swatch" is saturated and another product
has to be manufactured, the standard units can quickly be assembled to a
new machine. The application of light, modular slide units is the
premise for the production of variable precision machines for use in
exterior plants and sites.

Up to a short time ago gas and water turbines were brought back to the
work plant when a recondition was due. Their main parts had to be
changed completely; that led to enormous costs.

Nowadays industrialized as well as developping countries can't waste
money any more. For the reconditon of gas turbines, water turbines and
similar rotary machines, grinding and milling machines, which can be

Fig 2o Production maschine
for contact linses

Fig 2o

280

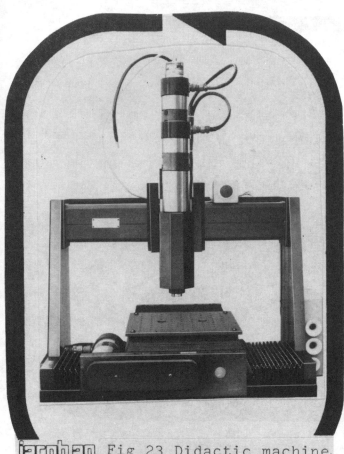

Fig 23 Didactic machine.

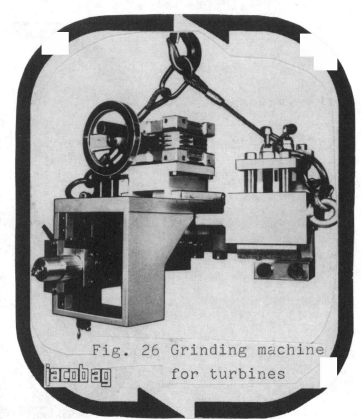

Fig. 26 Grinding machine
for turbines

demounted in their elements were
concipated. Like that, these
machines can be carried; the object,
which is to be worked , serves as
supporting table; the machines
consist only of the elements
necessary for the overhauling. As an
example this grinding machine: With
a grinding diameter of 800 - 1400 mm
and a length of 3000 mm it works
turbine blade heads in stators with
a truth of running of 0,025 mm.(Fig.
26) The grinding machine is placed

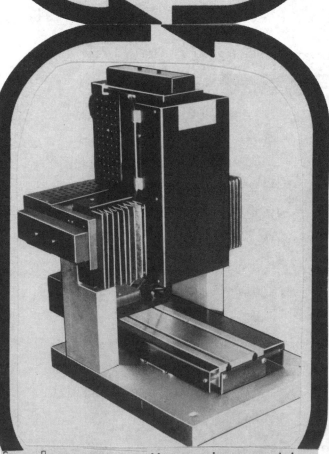

Fig 25 Measuring machine
for watches

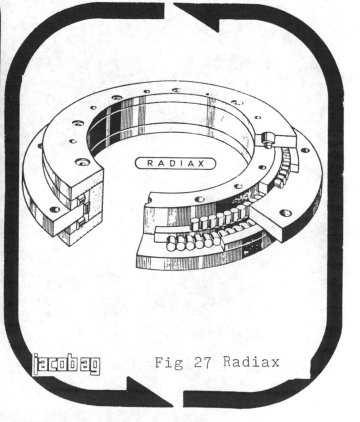

RADIAX

Fig 27 Radiax

into the opened turbine instead of the rotor. After reconvertion, which takes only 10 minutes, it is possible to grind stators as well. This application demonstrates a clear construction by means of modular slides.

These machines are working successfully worldwide.

Fig. 22 Scratch tester

A last advantage must not be unmentioned: The modular construction is interesting for all those countries, which are in the middle of industrial development: That means, they construct their own machines and buy only the elements, they can't produce themselves. With these elements and structures produced in their proper countries (cast iron, steel, concrete polymer) they can manufacture modern machines for their own requirement and export.

In this sense modular mechanical engineering makes a considerable contribution to the economic aid for developing countries.

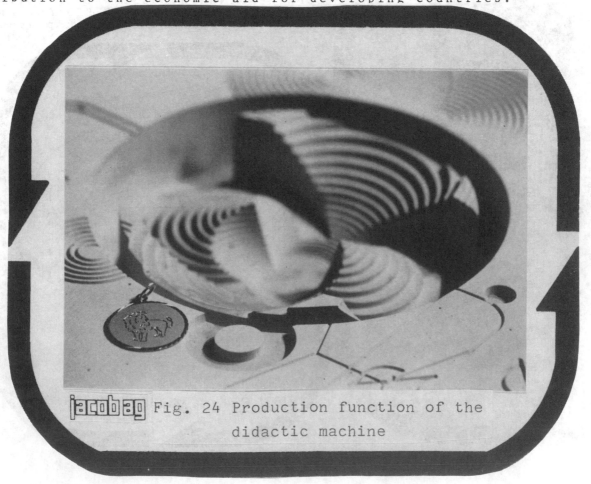
Fig. 24 Production function of the didactic machine

LATE PAPER

Low Cost Fixture for the Assembly of Hi-tech Product

Lennie E.N.LIM
Nanyang Technological Institute
Singapore

Abstract

This paper describes how low cost fixture can be made by using the vacuum themoforming technique in plastic. Design considerations on how to overcome the inherent limitations of this technique is presented. An example of a successful fixture for the assembly of a door mechanism in a computer disk drive, made by this method, is also discussed.

Introduction

The thought of manufacturing hi-tech product would immediately conjures up a need for a highly automated production line. This is not necessarily the case especially for products which are faced with rapid technological advances. Manufacturing such products cannot be held back until a highly automated production line is designed and installed. Furthermore, manufacturers are not prepared to invest so extensively in capital equipment for a product which is undergoing rapid technological changes. As such, these products are manufactured with semi-automated lines where fixtures of various designs are very widely used. Depending on the nature of the manufacturing processes, the number of any particular fixture required usually range from 10 to 1000 pieces. This quantity is not a very economical figure to work with and, therefore, a means of producing this limited quantity economically and expeditiously was investigated.

Costs of Fixture

The relative costs of fabricating fixtures using the various techniques is estimated and summarised in Fig.1.

It is evident from Fig.1 that, for a quantity of 10 to 1000 pieces, the relative cost of fixture fabrication using the plastic thermoforming technique is the most economical.

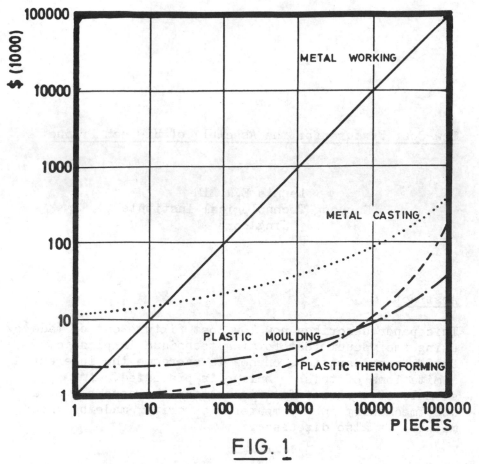

FIG. 1

Relative Costs of Fixture Fabrication

Thermoforming

Thermoforming is commonly used in packaging products, refrigerator liners, machine covers, light fitting and on a wide variety of thermoplastic materials (Ref 1). Its use as a technique to fabricate fixture has been hindered by its inherent limitations(Ref 2). However, its economic saving as well as its expeditious advantage over other fabrication techniques have prompted the present work. The limitations of this technique are overcome, to a great extent, by the mould design.

Thermoforming Mould

A male mould is preferred over that of a female mould because a better accuraccy is achieved as the stretching of the plastic sheet and hence its wall thickness, which can be substantial, need not have to be compensated. This also means that different material thicknesses may be used on the same mould without any modification which can be most advantageous. The bigger draft angle of 5° needed for the male mould, however, is used to advantage as the finished fixture will facilitate easier insertion of parts when used in the assembly line. The male mould also has the advantage of lower cost. (Ref 3).

Fig. 2 shows a typical fixture with radiused recesses, shown magnified for clarity. Generous allowances, in excess of 2 times the material thickness, are given to radii of bends and corners to facilitate easier and more accurate forming. The rigidity of the fixture may be increased by using thicker material or incorporating ribs in the fixture design.

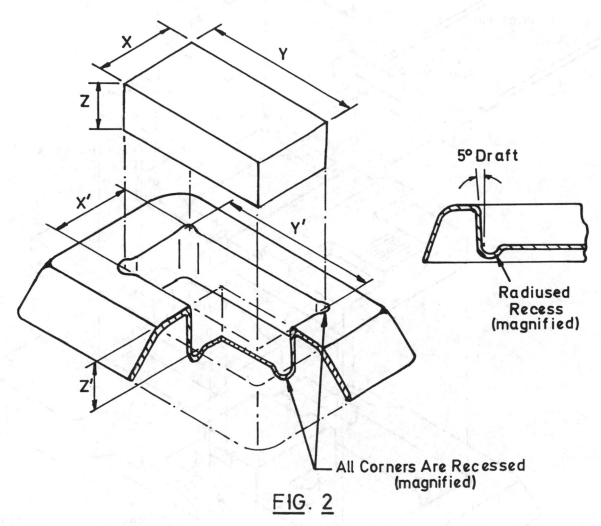

5° Draft

Radiused
Recess
(magnified)

All Corners Are Recessed
(magnified)

FIG. 2

Typical Fixture with Radiused Recesses

Example of Thermoformed Fixture

An example of a thermoformed fixture using 1.5 mm thick PVC is illustrated in Fig. 3.

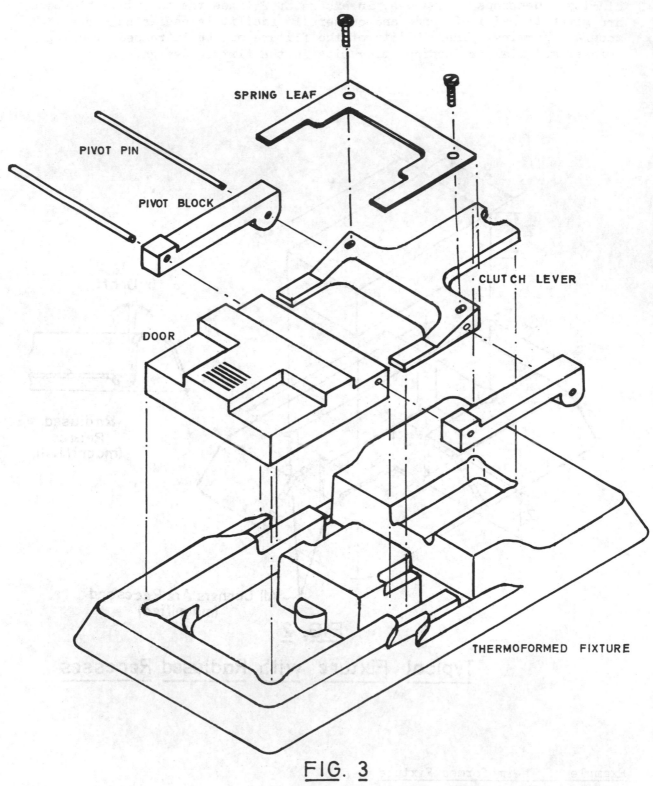

SPRING LEAF

PIVOT PIN

PIVOT BLOCK

CLUTCH LEVER

DOOR

THERMOFORMED FIXTURE

FIG. 3

Thermoformed Fixture for Door Mechanism

The fixture is used in the assembly of a door mechanism in a computer disk drive. The fixture is designed in such a way that it can provide the following functions :

a. A mean to locate the clutch lever and spring leaf accurately for them to be fastened by the 2 screws.

b. A further mean to locate the 2 pivot blocks and the door accurately for the 2 pivot pins to be inserted.

c. A mean to transport the "loosely-assembled" door assembly in one unit to the next assembly station.

d. A mean for easy removal of the "loosely-assembled" door assembly from the fixture and loading it on to the main body for final fixing.

For simplicity, Fig. 3 shows only one cavity. In practise, the actual fixture consists of 6 cavities with reinforcing ribs built into the design.

The use of the fixture has improved productivity in the door assembly by 200%.

Conclusion

Thermoforming can be used as a low cost and expeditious technique to make fixtures which are sufficiently accurate for light assembly work, especially in the electronic industries. Depending on the quantity of fixtures required, fixtures fabricated by this technique can cost from as low as 10% to 50% to that of fixtures fabricated by other conventional methods. Taking into account that the semi-automated assembly line would need numerous fixtures, the use of low cost fixtures using thermoforming can yield substantial savings and therefore profits.

Acknowledgement

The author would like to thank Nanyang Technological Institute for the use of its facilities and Wearnes Technology (Pte) Ltd for the permission to use the example presented in this paper. The work was carried out mainly at Wearnes Technology (Pte) Ltd for which the author is a consultant.

References

1. Brown R.L.E : "Design & Manufacturing of Plastic Parts" John Wiley, 1980

2. John D.Beadle, Ed. : "Plastic Forming" Macmillan, 1971

3. Milby, R.V. : "Plastic Technology" McGraw Hill, 1973

NEW AUTOMATED TECHNOLOGIES FOR THE PRODUCTION OF WELDED COMPONENTS

CESARE PANZERI, ANSALDO S.P.A., MILANO (ITALIA)

Memory proves that the automation of the whole welding cycle is conveniently reached with the new system fully self-adaptable to the forms of the joint or the robots developed for this purpose by author, conceived for creating a new "managed" factory architecture with an updated software, "arranged" according to flexibly production lines. The new production system based upon the multivariable control of the kinematic conditions, welding parameters and quality indices achieves all the advantages of productivity, welding quality flexibility of use, ease of programming and, above all, the solution of the exacting ambient problems.

STATE AND PROBLEMS OF THE ART

In its present state, the continuous joint welding installations con= sist of welding positions equipped for the use of manipulators fitted with welding heads requiring, though employing arc controls for con= tinuous welding, the presence of the operator in situ also on prehea= ted and difficult to reach pieces.

In fact, the manual guidance and groove tracking systems require the operator's presence for controlling, at each pass, the forward feed of the electrode to ensure a congruent welding bead deposition.

The welding joints normally encountered have the following characte= ristics:

. Joints of structural steel work pieces, also from a same lot, rare= ly satisfying precise geometric tolerances;

. Unique joints, also joints always differing one from the other, be= cause subject to deviations during the welding or preheating steps;

. Difficulty foreseeable, in a precise manner, evolution of the bead deposit shape.

In a consideration whereof, the "geometrical joint singularity" does not favour the use of guidance systems type "numerical control" or "self learning or preventive programming robot", difficult to use be= cause requiring, at each pass, the exact definition of the shape of the joint to be welded.

At the same time, the joint shape would have to be characterized by that constancy and repeatibility required for batch production.

The above applies not only to the continuous welding technology but also, in general, to all the other operations of welded structures construction cycles, above all the grinding operations carried out today with the operator in situ and manually. These operations, span= ning the whole range from joint preparation to back gouging and sur= face finishing or polishing with abrasive band as well as levelling also on double-bended surfaces, had to be automated.

The above presuppositions indicate that the proper answer to the pro= blems is found in the full automation of the process, thus leading to a better definition of the installations with the following advantages:

. Outstanding productivity increase

. Outstanding quality increase

. Improvement of working and safety conditions.

RESOLUTORY METHODOLOGY AND NOTEWORTHY CHARACTERISTICS

"Robotization" offers itself as the most suitable interdisciplinary technique for solving the above technological problems, keeping in mind also the necessity of the integration between the exigencies of the designer user and the constructive possibilities of the automation designer. Multipurpose robots are, under many aspects, not adaptable to the operative characteristics of continuous welding as well as con= tinuous grinding. The best way is an re-examination of the fabrication process on the basis of the indispensable quality factors prescri= bed by existing rules and new techniques offered by electronics and informatics.

The development of the new technology regarding the design and con= struction of new original units and systems is characterized by the following aspects:

1. Welding process probes and logics self-adaptive to the shape of the

joint to be welded or ground.

2. Development of a new welding software for facilitating the interac=
   tive colloquium operator-machine of the operating units among them=
   selves and the unit of the central management.

3. A new architecture with the creation of new systems, integrated,
   and of new flexible lines for making welded structures.

SELF-ADAPTABILITY

"Self-adaptability to the shape of the joint" means that the electrode
equipped with probes and logic (also by means of the welding arc) auto=
matically tracks the joint while welding it, like a driver at the guide
of his car adapts himself automatically to the configuration of
the road (straight, curved, obstacles etc.).

It is, in addition, possible to weld in successive passes from the bot=
tom to the top double curved joints (longitudinal and crosswise) using
suitable probes reading the line and surface coordinates (joint width)
and local inclinations).

These read coordinates taken while welding and stored with refe
rence to the actual torch position, are processed for obtaining refe=
rence signals of the multivariable control system consenting the auto=
matic concomitant adjustment of the kinematic block (machine axes) and
of the welding block (arc parameters) with the previous introduction
of the software representing the welding model.

The self-adaptive guidance unit, besides ensuring the absolute accuracy
of the bead deposits from the bottom to the top of the joint (with
storing of the position for accidental stops, manual steps or emergen=
cies) is fully interfaced with the other units for forming and integra=
ted modular system for the high-accuracy automatic control and diagnosis
of all the conditions and operative sequences of the welding process.

The self-adaptability of the electrode to the joint consents in addition
to the productivity and welding quality advantages also those of pro=
gramme flexibility and ease.

Two types of welding robots thus emerge as regards self-adaptivity:

1° Type : Fully self-adaptable robots not requiring any preventive self-
           learning programming on the piece to be welded. Destined for
           the welding of large and medium-sized bodies, structures etc.
           also with double-curved weld surfaces and relative long joints
           to be welded (> 1 meter), normally without sharp points.

2° Type : Programmable robots with self-adaptive local capacities regar=
           ding the joint. Destined for the welding of medium small bodies
           with sharp corners and discontinuities found in the
           joints to be welded.
           It is sufficient to programme or self-teach on the 1 st pie=
           ce of the batch; possible dimensional deviations with respect
           to other pieces will be compensated by the self-adaptive unit.

Figs. 1-2-3 show some welding systems of the 1st type, fully self-adap=
tive to the shape of the joint to be welded.

Fig. 4 shows a welding system of the 2nd type, i.e. self-adaptive pro=
grammable.

Fig. 5 shows a constant cutting force grinding unit with electronic
joint tracking system.

WELDING SOFTWARE

Within the context of designing a centralized welding process drive and control system, there arises the necessity of the precise definition of the methods by:

1) The organization of the actually existant data relating to the va= rious welding processes in a "data bank" obtained by the organic archiving of the said data over the appurtenant electronic termi= nals used for the interactive colloquium with the operator.
   The content of such a peripheral "data bank" can flow toward to or from the "central bank" suitable interconnected with the different welding stations.

2) The optimation of the application of various known welding proces= ses through the use and rationalized development of the ob tained data, possible through suitable preselection programmes "gui= ded", via terminal, welding parameters, regarding the kinematic aspe= cts, the shape and the arc generation.

3) The automatic correlation of the process parameter system with the evolution of the piece shape by using the automatic control of some variables included in the "data bank", discriminant for the choice of suitable parameters.

4) Introduction of automatic quality control factors during the weld= ing process as well as of basic parameters, while verifying any de= viation of the later with respect to the preset nominal values, si= gnalling of possible anomalies and their display on the terminal with recording of the event; if the parameter deviation exceeds the admitted qualitative standard, the working process if interrupted for allowing the verification of the anomaly.

5) Automatic diagnosis of the welding progress state, plant section defects and faulty apparatus, with clear indication thereof on the terminal by means of suitable messages and suspension as required of the process in course.

The accomplishment of the outlined innovations calls for electronic systems capable to fulfil the complex function of "flexible" manage= ment of the welding station as far as mechanic-kinematic aspects and arc-control are concerned.

These systems are assuring the possibility of allowing peripheral drive and control through suitably studied terminals for the standar= dization of the use of "welding software" (fig. 9), while still allow= ing the colloquium of the specific system with a higher level manage= ment unit designed for governing the different working stations.

The architeture of the system is required to resolve the serious envi= ronmental problems typical of welding in its functional and operative aspects.

ARCHITETURE

. The research of economic advantages deriving from the new producti= vity indices attainable from the new working qualities, facility of programming and the solution of serious environmental problems attainable with the new methods, leads to the definition of a new "ordered" factory architeture.

. Calls therefore for the generation of new integrated systems and flexible working lines, as identified for typical applications re= quired for the construction of various welded components.

It is possible to distinguish thus the following typologies:

1) LARGE WELDING SYSTEMS
   destined for the operations for the automatic welding of large-siz=
   ed pieces; the modular structure being designed for incorporating
   in addition to the automatically guided welding section, auxiliary
   sections for piece preparation and finishing (brushing, grinding,
   polishing) FIG. 6.
2) MEDIUM WELDING SYSTEMS
   compatible with medium-sized working processes; differing from the
   former in carrying structures, using base modules composing the auto=
   matic programmed and guided welding section (Fig. 7).
3) SMALL WELDING SYSTEMS
   for the automatic welding of small pieces, normally with very preci=
   se processes (MIG-TIG) (Fig. 8).

The above systems can be reciprocally associated for the working of
complex production components; in addition, they result to be inte=
grated with auxiliary systems allowing the flow continuity of the va
rious working phases provided for arriving at the finished product
such as:

4) PIECE PREPARATION / FINISHING SYSTEMS
   for the piece preparation assembling operations, grinding, surface
   finishing with abrasive band.
5) MANIPULATION SYSTEMS
   for the transfer of the pieces to the various processing stations
   and the local moving of the same.
6) SPECIAL MANIPULATION SYSTEMS
   for exacting manipulation steps under particularly difficult ambien
   tal conditions, comprising remote controlled versions with manual
   guide or programmable controls in the case of predetermined work=
   ing processes.

In view of their subsidiary nature, the systems 4, 5, 6, can be used
in different sectors in addition to the welded component sector (che=
mical nuclear etc. sectors).

With reference to the above-outlined systems and once defined the pro=
duction typology, the configuration of the welding line will be deter=
mined; however, the use will not be linked to a rigid product model,
but will remain highly flexible because of the modular constructive
and functional features of the various systems integrated therein.

The electronic structures, in particular, associated to the here con=
sidered systems, feature homogeneous characteristics regarding both
the constructive/hardware aspects (modularity, standardization of the
components, circuit operation reliability also under difficult indu=
strial conditions) and the intrinsic and operative programming/soft=
ware aspect.

Each system is accompanied by an electronic structure comprising the
following main parts:
- MAIN MANAGEMENT SECTION
- MECHANICAL-KINEMATIC OPERATION CONTROL SECTION
- ARC FUNCTIONING CONTROL SECTION (Specification of the welding sys=
  tems)
- INTERNAL INTERFACES
- INTERACTIVE COLLOQUIUM "TERMINAL" SECTION
- EXTERNAL INTERFACE WITH UPPER-LEVEL MANAGEMENT CENTRE

The circuit architecture comprises microprocessor configurations with given functions (also multi-distributed ones).

The interactive colloquium "terminal" allows the locally guided programming of all the specific working parameters, carrying-out controls, storing carried out programmes in a resident store with the possibility of recalling them for ripetitive working processes and the indication of diagnostic messages (Fig. 9). The external interface with the UPPER-LEVEL MANAGEMENT CENTRE allows supervision steps by means of the acquired knowledge of the state of the peripheral system, transfer of the inherent data relating to the programming/software of the working process (welding or other), synchronization of the working phases to ensure the optimation of the production flow.

The possibility of modular assembling of the flexible working lines based upon the use of standardized modular systems together with the flexibility of the development of a peripheral informatics both operative and for a central supervision, allows the progressive rational "arrangement" or architecture of the entire production process.

CONCLUSION

The sophisticated technology of the heavy components production for nuclear or conventional steam generation has made Ansaldo to proceed with the design, construction and sales of their own, original and patented systems for the welding and manipulation of welded components. The self-adjusted, multivariable and flexible control systems of advanced conception are the result of the application of electronic automation to component welding processes, not only for the heavy mechanical industry, but also for makers of medium-small welded structures.

Ansaldo, now already known on the international market for this new products typology born out of the twofold experience in the technological manufacturing sector and that of automation, intended to make use of this synergy for regenerating, thanks to the creation of the "new flexible working systems" both the technological production structure and the likewise strategical of the new design and production area of the new or flexible modular working lines and robots.

In particular, the ANSALDO GROUP, in accordance with the strategical concepts of the larger international groups, tends to promote a real technological development by using the big "laboratory" of their own divisions; in fact, there, outstanding innovative products, like the automatic systems for the production of welded components, are studied and preindustrialized for the purpose of supplying an efficient consolidation of the technology as well as qualified references on the international level.

LITERATURE

C. PANZERI - "Tecnologie di fabbricazione di componenti pesanti per centrali nucleari" XXIII. Nuclear Congress, Rome 16th - 17th March 1978
- "Mechanisiertes Schweissen von Kernreaktorkomponenti" Industrie Anzeiger - 4th October 1978
- "Welding Centers: New philosophy for automation" Mechanical Industry 1979
- "Welding of Vessel Joints - Automatic Guidance and Control Systems for Continuous Welding Self-Adaptive to the Joint Shape" Italian Machinery and Equipment November 1982

FIG. 1 Triaxial guidance unit for the continuous welding self-adaptive to joint shape

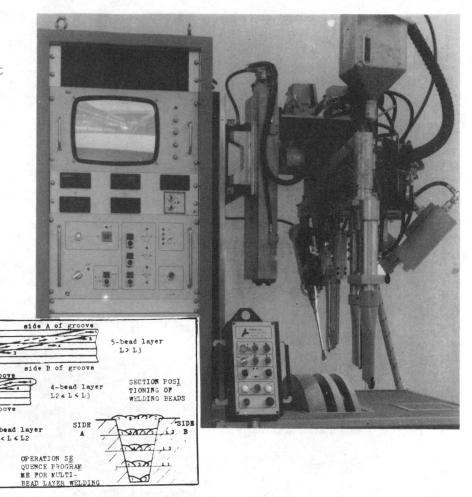

side A of groove

5-bead layer
L > L3

side B of groove

side A of groove

4-bead layer
L2 < L < L3

SECTION POSI
TIONING OF
WELDING BEADS

side B of groove

side A of groove

3-bead layer
L1 < L < L2

SIDE
A

SIDE
B

side B of groove

2-bead layer
L < L1

OPERATION SE
QUENCE PROGRAM
ME FOR MULTI-
BEAD LAYER WELDING

FIG. 2 Self-adaptive guidan= ce system for automatic stub pipe welding

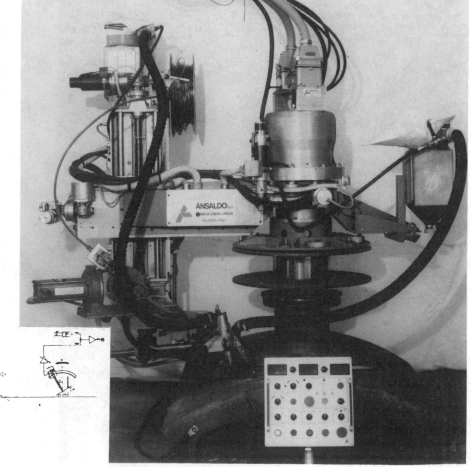

FIG. 3 Self-adaptive robot for the automatic continuous welding of longitudinal and girth joints and stub pipes (for the complete welding of girth and longitudinal joints and stub pipes always ensur<u>ing</u> the horizontality of the weld pool required for high productive processes also su<u>b</u> merged arc).

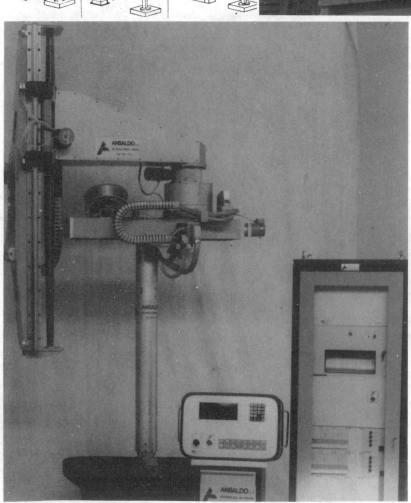

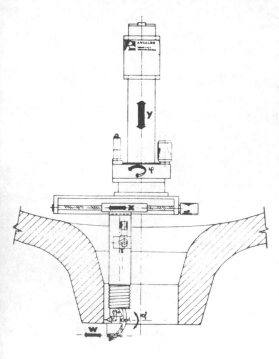

FIG. 4 Robot for stub-pipe cladding.

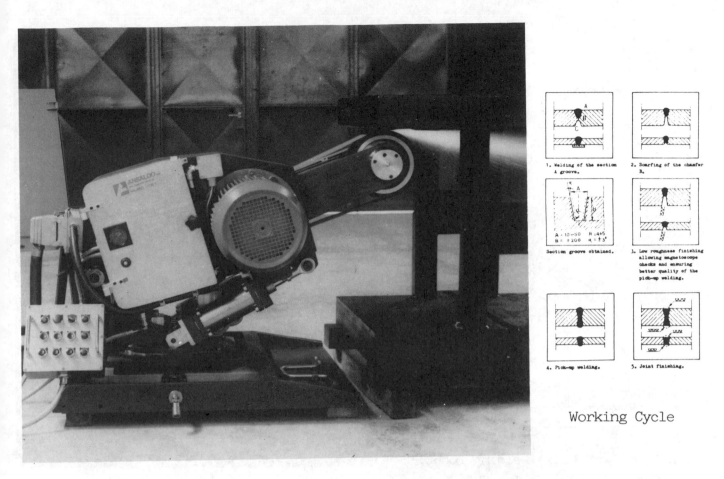

Working Cycle

FIG. 5 Automatic 3-Tool grinding unit with constant cutting force and electronic joint tracking.

FIG. 6 Integrated vessel welding center + Narrow Gap welding process;

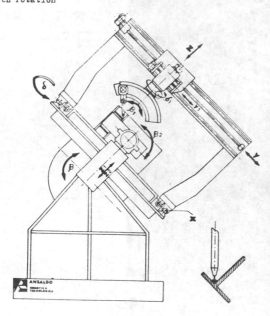

FIG. 7 Railway bogie welding robot: Programmable and self-adaptive to joint shape + auto-horizontality of weld pool.

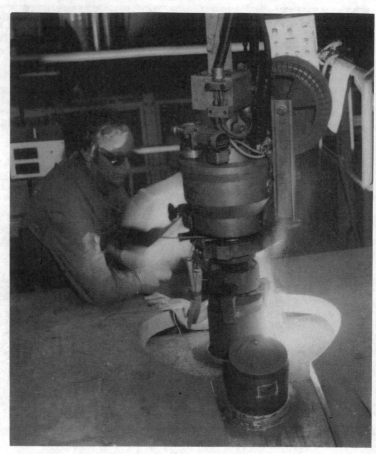

FIG. 8 Stub-pipe welding unit with self-adaptive electric arc.

FIG. 9 "Operative terminal" unit for drive and control with interactive colloquium

298